Science Starters

Natalie King

Contents

Page		Year 7	Year 8	Year 9
59	How fast?	F4		
60	Mass watchers	F5		
61	The Professor's friction	F6		
62	It's magic		F1	
63	Always, sometimes or never		F1	
64	That's nailed 'em		F3	
65	Shape up			F1
66	Just a moment			F2
67	Get the point			F3
68	Massive weights			F4
69	Size does matter			F4
	Energy			
70	Fuels forever	E1		
71	Alien energy	E2		
72	Electrical components	E2		
73	Warm nose, warm hands		E1	
74	A little light work		E2	
75	Cool vibrations		E3	
76	What a state		E4	
77	Energetic transfers			E1
78	You've got potential			E2
	Scientific enquiry			
79	Just an idea	SE1		
80	Dividing duckweed	SE2		
81	Ben's beans	SE3		
82	Know your tools	SE4		
83	Millie's measurements	SE5		
84	What grows in Emily's lawn?	SE6		
85	Graphic graphs	SE7		
86	What does it all mean?	SE7		
87	Solid as a rock	SE8		
88	New ideas		SE1	
89	In hot water		SE2	

About the authors

Natalie King is an experienced science teacher and is currently the head of science in a large city comprehensive school. She is a biologist.

Sarah Wyatt currently works as assistant head of science in a large city comprehensive school. She is an experienced science teacher and has worked in a variety of different schools and colleges of FE, and has worked in industry for many years too. She is a physical chemist.

Tom Kendall currently teaches science in a large city comprehensive school. He has also taught abroad as far afield as India. He is an ecologist.

Acknowledgements

We have all been fortunate to work with many enthusiastic and inspiring science teachers, and owe thanks to all of them.

Thanks go out to all our friends and family who have supported us along the way, especially to Delroy King for his constant support and encouragement throughout this project. Thanks also go to Rev. Adrian Heath and Veronica Heath for help and support. Finally, we would like to thank Emily and Joseph Hatton, David and Assa Kenall and Emily and Peter North who have been a great help too.

Introduction

Introducing starters

The aim of a starter activity is to get the students thinking. It is a 'warm up' for learning. It should be pacy and engage and challenge the students. Students should get into a routine of arriving at their lesson and getting straight on with a starter activity. They should expect this every lesson.

As endorsed by the Key Stage 3 Framework for Teaching Science, lessons should begin with a 10-minute starter activity, followed by the main part of the lesson and finally a plenary to check what learning has taken place and to help with future lesson planning.

The starter can be in many forms as long as it engages the students' thinking. It can be a written or an oral activity. Students may work individually, in pairs or as a group depending on the class.

This book contains a collection of starters that cover a range of teaching and learning activities. Very few resources are needed, which makes each activity easy to use in any teaching environment.

Target audience

Busy science teachers, experienced or NQT, who want inspiration and ideas to make the start of their lessons interesting, fun and pacy will find this book very useful.

This book is also suitable for non-specialist or cover/supply teachers who can use it as a source of ideas to engage the students in learning right at the start of every lesson. Often, cover teachers arrive at lessons in need of something to get the class settled and focused before the cover work arrives. These activities are ideal for this situation; with answers provided for each activity the non-specialist can feel more confident.

Any science department with recruitment and retention problems will find this book invaluable for ideas for setting cover lessons that are interesting and enjoyable. In fact, this book is suitable for any teacher who wants to bring fun and enjoyment as well as a swift start to their lessons with little time spent planning.

Using starters

Each starter is designed to focus on one framework objective, providing a focus to the start of the lesson.

You may want to start the lesson with a starter that relates to the main content of the lesson to get the students thinking about what they will learn in the lesson. Alternatively, you may want to use a starter activity which relates to a previous lesson to recap and refresh what has already been learnt before you move on to new concepts.

These starters are very flexible. They can also be used as plenary activities to check the students' understanding at the end of a lesson. There may even be some activities that you want to use within the main part of the lesson.

In conclusion, these starters are meant to complement your teaching of Key Stage 3 science, making sure you are fully embracing the Key Stage 3 National Strategy, and will hopefully inject some fun into your lessons.

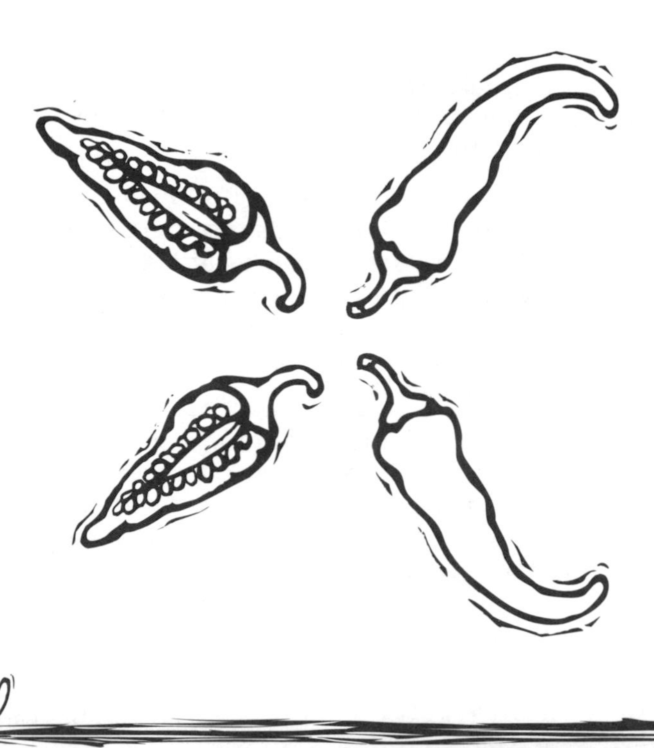

How to use this book

Contents grid

For simplicity, each starter has been assigned to one specific year group and one objective only. However, they are all very flexible and can in reality be used for any Key Stage 3 year group and adapted to your students' abilities.

Objectives and aims

Each starter activity has been matched to one Key Stage 3 science framework objective. Most starter activities have also been given a QCA reference to show where they match the QCA Scheme of Work. There is also an aim for each starter, which explains exactly what the activity is designed to achieve. However, the overall aim of these starters is for students to love learning about science and enjoy their science lessons. They will get the students thinking and motivated right from the very start of each lesson. They are also intended to remove some of the workload associated with planning lessons to meet the framework objectives.

Resources

Most of these starters require nothing more than a pen and paper plus an OHP or board. Activities have been chosen that require no photocopying or use of textbooks.

Activities

The activities are very clear and easy to follow, even for the non-specialist teacher. At most, they require minimal preparation on the OHP/board before the class arrives. Where this is necessary, a prepared OHT can be used again and again, eliminating the need for future preparation.

Differentiation

The main activity is aimed at the majority of students. The up arrow (⇑) represents ways to make the activity more challenging for the more able student. The down arrow (⇓) represents an easier version of the activity for the less able student. In some cases, these differentiations can be used as alternative activities in themselves or even as homework ideas.

Many science teachers have a great selection of starter activities that they know work in the classroom. However, they are often too busy to share these ideas with colleagues. From time to time, many of us need ideas to inspire our teaching and motivate the students. This book is intended as a bank of starter ideas that can be dipped in and out of to provide inspiration and complement our teaching.

The starters in this book have been tried and tested with a range of classes and they really do work. Don't be afraid to adapt the ideas for your own classes.

Enjoy!

Natalie King, Sarah Wyatt, Tom Kendall

Celebrating cells

Objective covered

QCA 7A. ***Framework C1.***

Aim

To recall the similarities and differences between plant and animal cells. To be able to recognise a plant cell and an animal cell.

Activity

- Draw the two diagrams below, without labels, on the board.

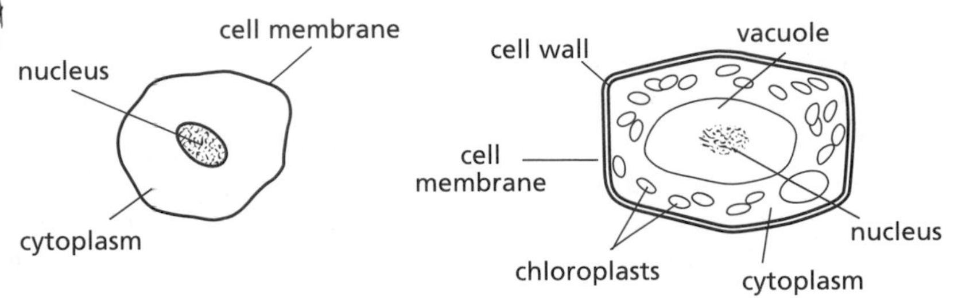

- Ask the students to look at the two drawings of a simple animal and plant cell. They should then draw both diagrams and add the following labels:

cell wall	cell membrane	nucleus
cytoplasm	vacuole	chloroplasts

Differentiation

- ⇑ 'Which cell is a plant cell, which is an animal cell and why?'
- ⇑ 'Why do you think plants don't need skeletons to keep them upright? What special feature of their cells gives them support?'
- ⇑ For a more able class you could scramble the label words up. The students then have to unscramble the words (5 minutes) before labelling the diagrams.
- ⇓ The first letter of each label can be given to help.
- ⇓ 'Animal cells have a blob-like shape, whereas plant cells have a more rigid box-like shape. Which cell is which?'

Organise the organisms

Objective covered

QCA 7A. ***Framework C2.***

Aim

To recall that some organisms are single-celled, while others are multi-celled.

Activity

- Write the names of these organisms on the board:

bacteria	yeast	elephant	daisy
flu virus	maggot	limpet	worm
protozoa	amoeba	*salmonella* bacteria	

- Ask the students to put the organisms into two categories: single-celled or multi-celled. Warn them that there is a hoax which does not fit. They should find the hoax and state the reason for their choice.

Answers

Single-celled: bacteria, yeast, *salmonella* bacteria, protozoa, amoeba
Multi-celled: elephant, daisy, maggot, limpet, worm
The hoax is flu virus, as it is not a living thing and is not made of cells. Instead it is genetic information surrounded by a protein coat.

Differentiation

- ⇑ Ask the students to think of an extra organism they could add to each category. Alternatively, you could add more organisms to the list for them to sort.

- ⇓ Only give half the names for students to sort. Alternatively, you could give them pictures of the organisms to sort into the two categories.

Tree rings

Objective covered

QCA 7A. ***Framework C3.***

Aim

To recall that growth means an increase in size and in the number of cells. To be able to draw a line graph and interpret data.

Activity

- Explain that trees grow steadily each year throughout their life. You can tell the age of trees by measuring their girth.
- Display these data:

Age of tree (years)	Girth (cm)
10	45
20	85
30	140
40	180
50	210

- Ask the students to draw a line graph.
- From their graphs, ask them to answer these three questions:

  1 What girth would you expect a 35-year-old tree to have?
  2 Is the tree still growing even though it's 50 years old?
  3 What do you think its girth will be at 60 years old?

Answers

1 160 cm
2 Yes
3 230–40 cm

Differentiation

- ⇑ 'Why is it more accurate to measure girth than width?' (It is easier to measure.)
- ⇓ Students can be given graph paper with axes pre-drawn. Bar graphs may be easier for them to draw.

Very special cells

Objective covered

QCA 7A and 7B. ***Framework C4.***

Aim

To recall that specialised cells have special features which help them to do their job.

Activity

- Display the following diagrams:

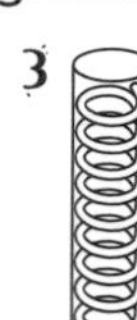

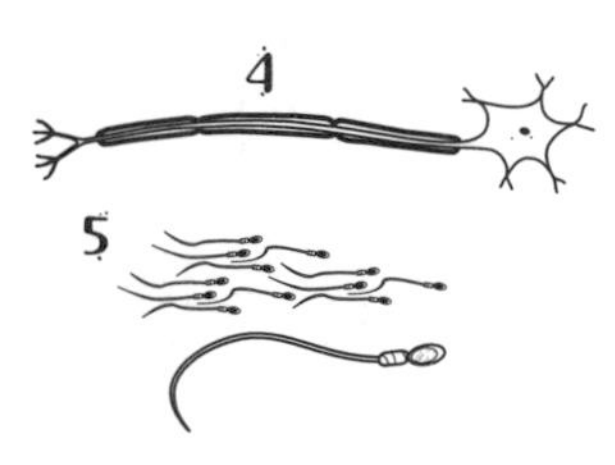

- Ask the students to copy the diagrams of specialised cells into their books.
- Next, write these jobs/descriptions on the board:
    - *a* No nucleus, disc-shaped. Carry oxygen.
    - *b* Have a tail to swim to the egg. The head contains the nucleus.
    - *c* Wire-like. Carry electrical messages around the body.
    - *d* Tube-like. Carry water through the plant.
    - *e* Blob-shaped, with irregular nucleus. Engulf microbes.
- Tell the students to match the jobs to their diagrams.

Answers

1 e 2 a 3 d 4 c 5 b

Differentiation

- ⇑ Ask the students why red blood cells do not have a nucleus.
- ⇓ Give drawings of the cells to the students; get them to match names and jobs to cells.

Who am I?

YEAR 7

Objective covered

QCA 7A and 7B. ***Framework C4.***

Aim

To recall that specialised cells have special features which help them to do their job.

Activity

- Ask the students to name and draw these specialised cells:

  1. Who am I? I am disc-shaped with no nucleus. I squeeze in and out of tight places.
  2. Who am I? I am wire-like. I carry important messages at high speed. I am long.
  3. Who am I? I am very large and I only have half the genetic information in my nucleus. I am important to life.
  4. Who am I? I have a tail to swim. I have half the genetic information in my nucleus.

Answers

1 Red blood cell 2 Nerve cell 3 Egg cell 4 Sperm cell

For diagrams see 'Very special cells', page 4.

Differentiation

- ⇑ Add 'Who am I? I carry water, I am tube-like.' (xylem cell)
- ⇓ Give the students pictures to help them or get them to draw cartoon cells to match descriptions.

Tissues and organs

Objective covered

QCA 7A. ***Framework C5.***

Aim

To recall what a tissue and an organ are and be able to give examples.

Activity

- Display these words on the board:

lungs	brain	muscle	retina	skin
cartilage	heart	fat cells	nerves	liver
pancreas	stomach	stomach lining		

- Tell the students to get into pairs. Ask them to sort the words into two groups: tissues or organs.
- Ask the students to tell their partner what an organ is and what a tissue is.

Answers

Organs: lungs, brain, skin, heart, liver, pancreas, stomach
Tissues: muscle, retina, cartilage, fat cells, nerves, stomach lining

Similar cells working together are called tissues.
Different tissues working together are called organs.

Differentiation

- ⇑ Ask the students what the organs in the list do, or to make a list of plant organs.
- ⇓ Give students the definitions of organs and tissues to help them sort the words. For very low ability students, show them pictures of the organs to help with the words.

Lookalikes

Objective covered

QCA 7B. ***Framework C6.***

Aim

To recall that during sexual reproduction the offspring are not identical but similar because there is a sharing of genes.

Activity

- Ask the students to think of someone they know who looks like one of their parents. They should list all the ways they are similar to their mum or dad. They should also list the ways they are different.
- 'Now think of the Queen. List all the ways the Princess Royal (Anne) is similar to her and all the ways she is different.'
- 'Now think of David Beckham. How is his son Brooklyn the same as him, and how is he different?'

Answers

You may discuss similarities and differences in hair colour, eye colour or shape, height, nose, forehead, texture of hair, temperament, weight, mouth and lips, intelligence and so on.

Differentiation

- ⇑ 'Which of the factors discussed are not just down to genetics. Which can be influenced by environment?' (weight, temperament, IQ)
- ⇓ Pictures of people from the same family may help focus the students' ideas, for example David and Brooklyn Beckham.

Joining up

Objective covered

QCA 7B. ***Framework C6.***

Aim

To recall what fertilisation means.

Activity

- Write these anagrams of key words on the board:

  nitoasireftil euslcun geg mrpse gnijoni xse

  1 Ask the students to unscramble the key words.

  2 The students should then write two sentences that include all the key words.

Answers

1 fertilisation, nucleus, egg, sperm, joining, sex
2 For example: Fertilisation is the joining of the nucleus of an egg with a nucleus of a sperm. This may happen during sex.

Differentiation

- ⇑ 'Why are the resulting offspring not identical to the parents?'
- ⇓ Give the students the unscrambled key words. They can then make the sentences.

What's in a food?

Objective covered

QCA 8A. ***Framework C1.***

Aim

To recall the seven parts to a healthy diet and recall what the main nutritional groups are for. To be able to give examples of foods for each group.

Activity

- Draw this table on the board:

Carbohydrates	Fats	Proteins

- Call these foods out one at a time and ask students which column of the table they should go in: bread, meat, eggs, butter, pasta, rice, milk, cheese, fish, potatoes, beans, cereal, oil, cream.
- Ask the students to write a sentence to explain what each of the food groups is for.
- Ask the students what else is needed for a healthy diet.

Answers

- Carbohydrates: bread, pasta, rice, potatoes, cereal (beans could go here too)
  Fats: butter, oil, cream (cheese has a lot of fat too)
  Proteins: meat, eggs, milk, cheese, fish, beans (some in cream)
- Carbohydrates are for energy. Proteins are for growth and repair. Fats are for energy, storage and to protect organs.
- Fibre, water, vitamins and minerals are also needed for a healthy diet.

Differentiation

- ⇑ Add more foods to the list. Ask which foods can appear in more than one column. For example, peanut butter: fat and protein. Ask the students to write down what they ate yesterday and then add the foods to the appropriate columns in the table.
- ⇓ Give just two foods from each group to sort.

Help the dietician

Objective covered

QCA 8A. ***Framework C1.***

Aim

To recall the seven parts to a healthy diet and recall what the main nutritional groups are for. To be able to give examples of foods for each group.

Activity

- Display these sentences:
  1. Matthew has broken his arm.
  2. Tracey is breast-feeding her newborn baby.
  3. Issac is training for a marathon.
  4. Adrian is trying to lose a bit of weight.
  5. Lily is recovering from a heart attack.
  6. Portia is a growing 5-year-old.
  7. Emily has just had the flu.
- Explain that Del is a hospital dietician and that the people on the board are some of his patients. Ask the class to help Del by saying what food groups he should tell his patients to eat more or less of.

Answers

1 More protein 2 More protein 3 More carbohydrates 4 Less fat
5 Less fat 6 More protein 7 More protein

Differentiation

- ⇑ Ask the students what vitamin or mineral is missing to cause these diseases: scurvy, anaemia and rickets.
- ⇓ Display sentences 1, 3 and 4 only.

Chemical scissors

Objective covered

QCA 8A. ***Framework C2.***

Aim

To recall why some molecules need to be broken down before the body can use them. To recall the role of enzymes in digestion.

Resources

Scissors and paper.

Activity

- Draw a starch molecule on the board and ask the students to copy it onto paper.

  ●-●-●-●-●-●-●-●-●-●-●-●-●-●-●

- Ask the students why it is hard for the starch molecule to pass through the gut wall and into the blood where it is needed.
- The students should now cut the starch molecule up, making cuts between each of the circles.
- 'Is it easier for the cut-up/digested starch to pass through the gut wall and into the blood? Why?'
- Ask the students if they know what does the 'cutting up' in their digestive systems.

Answers

- The starch molecule is too large to pass through the gut wall.
- The starch molecule has been digested into sugar molecules which are small enough to fit through the gut wall.
- Enzymes

Differentiation

- ⇑ 'Can you name the enzyme that digests starch?' (amylase) 'Why do washing powders contain enzymes?'

Freddy Food's journey

Objective covered/aim

QCA 8A. ***Framework C3.***

Activity

- Tell the students that they are going to investigate the journey any item of food takes through the digestive system. Let them know that one useful memory aid is to personify this item, as Freddy Food.
- Display these names of parts of the digestive system:

large intestine	anus	mouth
gullet (oesophagus)	stomach	small intestine

- Ask the students for the order in which Freddy passes the parts.
- Finally ask these questions:

1. Where do Freddy's broken-down useful parts pass into the blood?
2. At which part is excess water reabsorbed?
3. Which parts are involved in breaking him down?
4. Which parts contain enzymes?

Answers

Correct order: mouth, gullet, stomach, small intestine, large intestine, anus.

1 Small intestine
2 Large intestine
3 Mouth, stomach and first part of small intestine
4 Mouth, stomach and small intestine

Differentiation

- ⇑ Ask the students to write a detailed paragraph using the above words to explain how Freddy Food gets from the plate, into the blood and to the cells that need his energy. They should name the enzymes involved.
- ⇓ Ask the students to draw cartoons to show what is happening at each of the parts.

Myrtle's marathon

Objective covered/aim

QCA 8B. ***Framework C4.***

Activity

- Tell the students that Myrtle is running a marathon for the first time. She will be running for quite a few hours so her muscles need lots of energy. Her friends notice some changes in her during the race:

  1 She starts to look red.
  2 Her heartbeat increases.
  3 Her breathing rate increases.
  4 She needs to drink more.
  5 She loses water in sweat.

- Get the students to explain the changes and then ask them:

  6 Should her friends call an ambulance?
  7 What is the process that converts glucose and oxygen into energy for cells?

Answers

1 Blood is directed to the surface of the skin to help her cool down.
2 Her heart pumps blood faster to supply the muscles with more oxygen and glucose needed for respiration.
3 Her breathing rate increases to supply the muscles with more oxygen needed for respiration.
4 Water is needed to replace the water lost in sweating.
5 Sweat evaporates, drawing heat energy away from her body to cool her down.
6 No, the changes are normal.
7 Respiration. (Respiration is not breathing.)

Differentiation

- ⇑ The students should write the word equation for respiration.
- ⇓ Ask the students to draw the runner and label the changes her friends notice.

Microbes everywhere

Objective covered

QCA 8C. *Framework C5.*

Aim

To recall the three types of microbe. To recall they are so small you need a microscope to see them. To recall that some microbes are useful and some are harmful.

Activity

- Write these anagrams of key words on the board:

  cabiaret unfig sivur

- Get the students to unscramble the key words.
- Next ask these questions about the key words:
  1. Give an example for each of how they can be useful or harmful to us.
  2. Which is the smallest?
  3. Which is smaller than a human cell?

Answers

Bacteria, fungi, virus

1. Bacteria: Useful – making antibiotics and drugs; live in your gut to aid digestion; yoghurt. Harmful – can cause food poisoning and disease.
   Fungi: Useful – as food, for example, mushrooms and quorn; yeast is used to make wine and bread. Harmful – athlete's foot.
   Virus: Useful – used in genetic engineering. Harmful – cause diseases like flu and HIV.
2. Virus
3. All of them

Differentiation

- ⇑ 'Which one is not living?' (virus) 'Explain why.'
- ⇓ Tell students the ways in which microbes can be useful or harmful. Get them to match each one to bacteria, fungi or virus.

Keeping microbes out

YEAR 8

Objective covered

QCA 8C. ***Framework C6.***

Aim

To recall some of the ways that the human body kills microbes and stops them from entering.

Resources

Poster/sugar paper.
Chunky pens.

Activity

- Draw a large outline of the human body on the board.
- Ask students to copy the diagram onto poster paper.
- In pairs, they should add labels to the diagram of all the ways the body destroys microbes or stops them from entering.

Answers

Labels should be drawn to the following areas:

- eyes – tears contain enzymes that kill microbes
- stomach – acid kills microbes
- lungs/breathing system – mucus traps microbes, cilia help waft them out
- intestines – get rid of microbes and toxins through diarrhoea
- skin – keeps out microbes
- vagina – low pH here stops many microbes growing.

Differentiation

- ⇑ Students can sketch organs on their diagram before adding labels.
- ⇓ Tell the students a few of the ways the body kills or keeps out microbes. They can then use this information to add to their diagram. Alternatively, let them work in a pair with a more able student.

This will hurt a little bit

Objective covered

QCA 8C. Framework C6.

Aim

To recall how immunisation improves immunity.

Resources

OHP and OHT, board or worksheet prepared with the following questions:

1. What vaccinations do we receive as a child?
2. Who pays for them? Where does the money come from?
3. Who administers them?
4. Do you know anyone of your age who has had mumps or measles? If not, why not?
5. What is in the vaccine you are given?
6. What do your white cells make in response?

Activity

- Ask the students to discuss the questions in pairs. One should write down their thoughts, the other should report back to the whole class.

Answers

1. Measles, mumps, rubella, polio, diphtheria and tetanus.
2. The government. The money comes through taxes.
3. Nurse, doctor or health visitor.
4. Probably not, as (almost) everyone has been vaccinated against these.
5. A dead or weakened form of the disease-causing microbe.
6. They make antibodies to destroy the microbe.

Differentiation

- ⇑ 'Why do antibodies stay in your blood after an infection? Why don't you get ill when you are vaccinated?'
- ⇓ 'Make a list of "jabs" you think you've had. Have you had any of these diseases? Why not?'

Burn baby burn

Objective covered

QCA 9B. ***Framework C1.***

Aim

To recall the similarities between respiration and combustion.

Activity

- Get the students to use their knowledge to answer these questions:
  1. When you walk down the road what fuel are you using?
  2. When you drive your car what fuel are you using?
  3. If you are put in a room without oxygen what will happen?
  4. What will happen if you stop oxygen from getting to a fire?
  5. What is the equation for respiration?
  6. What is the equation for burning a fuel?
  7. Can you see any similarities between the two equations?

Answers

1 Food/glucose
2 Petrol/diesel
3 You will die.
4 It will go out.
5 Glucose + oxygen → carbon dioxide + water + ENERGY
6 Fuel + oxygen → carbon dioxide + water + ENERGY
7 The two equations are nearly identical.

Differentiation

- ⇑ More able students may be able to write the equation in symbol form.

  $C_6H_{12}O_6 + 6O_2 \rightarrow 6CO_2 + 6H_2O$ + ENERGY

  $CH_4 + 2O_2 \rightarrow CO_2 + 2H_2O$ + ENERGY

- ⇓ Less able students may need to be given the equations with parts missing.

  Fuel + _______ → ________ _______ + water + E_____

Bad news

Objective covered

QCA 9B. ***Framework C2.***

Aim

To recall the effects of drugs on the body.

Activity

- Get the students to use their knowledge to answer these questions.
  1. List as many drugs as you can.
  2. Which of these drugs are legal?
  3. In order to affect you, which part of your body must drugs affect?
  4. Are any of the effects of drugs permanent?
  5. What do you think the long-term effect of drug use may be?

Answers

1. Could include alcohol, nicotine, heroin, LSD, ecstasy, marijuana and so on.
2. Alcohol, caffeine, nicotine
3. Brain and nervous system
4. Yes – permanent damage to liver caused by alcohol, death by heroin and so on.
5. Physical damage to body and nervous system, psychological damage, damage to society caused by robbery and so on.

Differentiation

- ⇑ More able students could consider the consequences of drugs on society.
- ⇓ Less able groups could write an anti-drugs rap song.

Keeping healthy

YEAR 9

Objective covered

QCA 9B. ***Framework C2.***

Aim

To recall that our bodies need to be looked after.

Resources

OHP and OHT, board or worksheet prepared with the following questions:

1 List as many parts of a car as you can.
2 What happens to the car if it gets a puncture?
3 What happens to the car if you put the wrong type of fuel in the tank?
4 List as many parts of the human body as you can.
5 What will happen to your body if you do not do enough exercise?
6 What will happen to your body if you eat the wrong types of food? (For example, too much fat.)

Activity

- Get the students to use their knowledge to answer the questions in pairs.
- Discuss answers as a group.

Differentiation

- ⇑ Ask the students to develop their answer to question 6 by discussing a healthy diet or analysing their own diet and lifestyle. 'What are the alternatives to burgers and chips?'
- ⇓ The students could draw pictures to describe a healthy lifestyle.

Who are you?

Objective covered

QCA 9A. ***Framework C3.***

Aim

To recall that we inherit our features from our parents.

Resources

OHP and OHT, board or worksheet prepared with the following questions:

1 List the features that you have inherited from your parents, for example your nose.
2 What are the male and female sex cells called?
3 What molecule carries coded instructions to make you?
4 What do we call the meeting of the male and female gametes?
5 Why are brothers and sisters from the same parents often so different from each other?

Activity

- Get the students to use their knowledge to answer the questions.

Answers

1 All features
2 Sperm and egg
3 DNA (deoxyribonucleic acid)
4 Fertilisation
5 They have a different set of genes.

Differentiation

- ⇑ Students could consider the nature/nurture argument for personality.
- ⇓ Ask the students to draw the features they inherited from each parent, for example their mother's nose.

Warren's cow

YEAR 9

Objective covered

QCA 9A. ***Framework C3.***

Aim

To recall that selective breeding is used to produce 'perfect' farm animals.

Activity

- Tell the students that Farmer William Warren is trying to breed the finest herd of cattle in the West. He has two bulls, one is big and muscular, the other is really skinny.
- The students should then answer these questions:
  1. Which bull should he breed from?
  2. Explain your answer to question 1.
  3. What qualities will he be looking for in his herd of prize-winning cows?
  4. What do we call the process of trying to make 'perfect' animals by artificially selecting mates?

Answers

1 The big and muscular one
2 This bull is more likely to pass on the genes for large muscular cows.
3 Size, fat content, taste of meat, lack of aggression and so on
4 Selective breeding

Differentiation

- ⇑ The students could consider the negative examples of selective breeding, for example: dogs – weak backs, poor eyes, hip problems.
- ⇓ Ask the students to draw idealised animals, for example a fat pig or a huge bull.

Light food

Objective covered

QCA 9C. ***Framework C4.***

Aim

To recall that plants use photosynthesis to make their food.

Activity

- Get the students to use their knowledge to answer these questions:
  1. Where do plants get their energy from?
  2. What other things do plants need to be able to grow?
  3. What do we call the process whereby plants use energy from the Sun to be able to grow?
  4. Write a word equation for photosynthesis.

Answers

1 The Sun
2 Carbon dioxide and water
3 Photosynthesis
4 Carbon dioxide + water $\xrightarrow[\text{chlorophyll}]{\text{sunlight}}$ glucose + oxygen

Differentiation

- ⇑ The students could use chemical symbols to write the equation for photosynthesis:

$$6CO_2 + 6H_2O \xrightarrow[\text{chlorophyll}]{\text{sunlight}} C_6H_{12}O_6 + 6O_2$$

- ⇓ Ask the students to draw a plant and show how the factors needed for photosynthesis enter the plant and how waste products leave the plant.

Perfect plants

Objective covered

QCA 9C. ***Framework C4.***

Aim

To recall how plants are adapted for photosynthesis.

Activity

- Get a student to sketch a plant on the board.
- Then get other students to label a main part each (leaves, stem, root, fruit, flower). They may need to add parts to the original sketch.
- Ask the students how the roots are adapted to collect as much water as possible. (They are very long, with root hairs to increase surface area.)
- Ask how the leaves are adapted to collect as much sunlight as possible. (They have a large surface area, are thin and arrange themselves to avoid shading each other.)
- Discuss other adaptations plants have made to make photosynthesis more efficient. For example, long stems to compete with other plants for sunlight.

Differentiation

- ⇑ Ask the students to plan an investigation to find out the total surface area of all the leaves on a tree.
- ⇓ Draw the plant on the board or a worksheet for the students to label.

Busy Lizzie

Objective covered

QCA 9C. *Framework C5.*

Aim

To distinguish between photosynthesis and respiration in plants, including the use of word equations.

Resources

OHP and OHT, board or worksheet prepared with the equations for photosynthesis and respiration:

$$\text{Carbon dioxide + water} \xrightarrow[\text{chlorophyll}]{\text{sunlight}} \text{glucose + oxygen}$$

$$\text{Glucose + oxygen} \longrightarrow \text{carbon dioxide + water + ENERGY}$$

Activity

- Get the students to use their knowledge to answer these questions:
  1. Can you see any links between these two equations?
  2. Where do plants get the energy to make glucose from?
  3. Which of the seven life processes uses the glucose produced by photosynthesis?
  4. When we eat an apple, the glucose is used to make energy in our cells by respiration. Where did the energy released by respiration originally come from?

Answers

1 The equations are nearly the opposite way around.
2 Sunlight 3 Respiration 4 Sunlight

Differentiation

- ⇑ Give the equations for photosynthesis and respiration in symbol form:

$$6CO_2 + 6H_2O \xrightarrow[\text{chlorophyll}]{\text{sunlight}} C_6H_{12}O_6 + 6O_2$$

$$C_6H_{12}O_6 + 6O_2 \longrightarrow 6CO_2 + 6H_2O + \text{ENERGY}$$

Sorting animals

Objective covered

QCA 7D. ***Framework I1.***

Aim

To recall features of the main groups of animals: mammals, fish, birds, amphibians and reptiles.

Activity

- Draw this table on the board:

Group	Organism	Reason
Mammal		
Fish		
Bird		
Amphibian		
Reptile		

- Ask the students to put the following organisms into the correct column and give a reason for their choice.

  monkey   stickleback   frog   sparrow   snake

Answers

Monkeys are mammals as they have fur, give milk to their young and give birth to live babies. Sticklebacks are fish as they live all the time in water, swim using fins, and have scales and gills. Frogs are amphibians as they live partly on land and partly in water. Frogs also return to the water to breed. Sparrows are birds as they have feathers and wings, and lay eggs. Snakes are reptiles as they have dry scaly skin and lay eggs on land.

Differentiation

- ⇑ Ask the students to place these organisms too: killer whale, dolphin, penguin, duck-billed platypus.
- ⇓ Ask the students to match the organisms to groups without giving reasons, or alternatively giving very simple ones.

Animal experiments

YEAR 7

Objective covered/aim

QCA 7D. ***Framework I1.***

Resources

OHP and OHT, board or worksheet prepared with one of the following statements:

- Doing experiments on animals is okay because it makes our world a safer place.
- Experiments on animals should be banned even if it means that we cannot ensure that our medicines will be safe.

Activity

- Allow the students five minutes to think of arguments to support both sides of the statement. They should record their arguments in a table:

The argument is right because	The argument is wrong because

- Students should then feed back their arguments to the whole class.
- Challenge students with counter arguments, for example:
  - Do you eat meat?
  - Have you ever been in hospital/taken medicine?
  - Do you use a shampoo? How many animals were used to test it?

Differentiation

- ⇑ Give the students an extra five minutes to prepare a two-minute speech on their point of view.
- ⇓ Less able students could concentrate on only one side of the argument (split the class in two).

Food webs all around

Objective covered

QCA 7C. ***Framework I2.***

Aim

To recall that animals do not eat just one type of food. Because of this, food chains link up to make food webs. This gives a better representation of what's going on in an ecosystem.

Activity

- Display this woodland food web on the board:

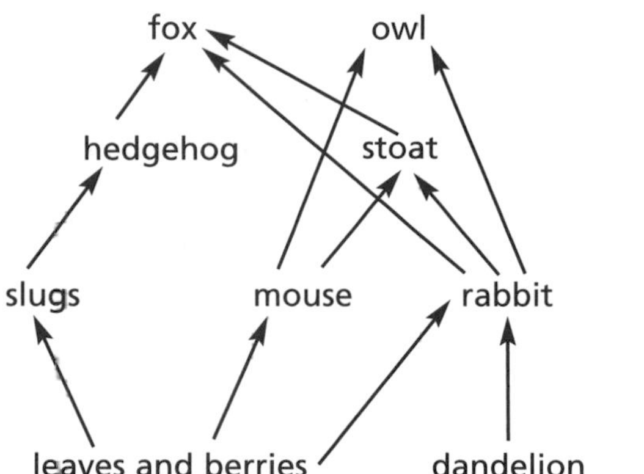

- Ask the students to write down as many food chains as they can from the web.

Answers

For example: dandelion → rabbit → fox

Differentiation

- ⇑ Ask further questions: 'How many carnivores are there in this web? How many herbivores? Name them. What do all food chains start with?'
- ⇓ Ask the students to find only three food chains from the web.

Surviving change

Objective covered

QCA 7C. *Framework I3.*

Aim

To recall how some organisms adapt to changing seasons. To recall how animals adapt the way they feed to survive. To recall why some organisms are better at surviving in different habitats.

Activity

- Write these organisms on the board:

goose	dormouse	frog	hedgehog
tulip	greenfly	snowshoe hare	oak tree

- Ask the students how each of the organisms survives the winter.
- Find out if any students know what koala bears or pandas eat. Ask them why this is not a good idea.

Answers

- Geese migrate to warmer countries. Dormice, frogs and hedgehogs hibernate. Tulips survive as bulbs. Greenfly die, but they lay eggs which survive; new greenfly hatch out in the spring. Snowshoe hares grow thicker coats. Oak trees lose their leaves.
- Koalas eat eucalyptus. Pandas eat bamboo. Eating only one type of food is not a good idea because if it runs out the organism will starve and die out.

Differentiation

- ⇑ 'Successful organisms can adapt to their changing environments and can survive in more than one habitat. Are humans successful organisms? Why?'
- ⇓ 'How do dogs, cats and tortoises survive the winter? Why do squirrels eat lots in autumn and gather lots of food?'

Unscramble the plants

Objective covered

Framework I1.

Aim

To recall the names and features of the main plant groups.

Activity

- Display these anagrams of plant groups on the board:

  sreifnoc   nrefs   somsse   wolfgnire stnpla

- Ask the students to unscramble the anagrams.
- Ask the class these questions, identifying the plant groups:
  1. Which one produces seeds in cones?
  2. Which one produces flowers or fruits?
  3. Which one grows on walls and damp places and has thin, tiny leaves?
  4. Which one grows big with strong roots and leaves, and makes spores?

Answers

conifers, ferns, mosses, flowering plants
1 Conifers   2 Flowering plants   3 Mosses   4 Ferns

Differentiation

- ⇑ Ask the students how the spores and seeds of the plants might be spread.
- ⇓ Just ask questions 1 – 4. You may give the students the unscrambled words.

Which habitat?

Objective covered

QCA 8D. ***Framework I2.***

Aim

To recall that habitats can change, and some reasons why.

Activity

- Write these words on the board, in two columns:

frog	desert
pike	wood
dandelion	stream
squirrel	pond
camel	path

- Get the students to match the living things to their environment.
- Ask them what things may cause changes to a woodland habitat.

Answers

- Frog–pond, pike–stream, dandelion–path, squirrel–wood, camel–desert
- Causes given by students may include: cutting down/clearing of trees; chemicals/herbicides may kill plants; disease may affect plants and trees; one type of organism may overgraze the area; population explosion; extremes of weather; natural disasters.

Differentiation

- ⇑ Ask the students how people are causing a change in habitats around the world.
- ⇓ 'Your house is part of your habitat. What does it give to you? Make a list of things that could change it.'

What rotters

Objective covered

QCA 9G. Framework I3.

Aim

To understand that living things eventually die and rot.
They also produce wastes, which rot.

Activity

- Display these words:

  cork rubber cotton glass vegetable peelings plastic
  wood paper perspex china apple

- Ask the students which of the materials displayed are biodegradable.
- Ask the students what happens to the minerals and compounds in the biodegradable materials.

Answers

- Cork, cotton, vegetable peelings, wood, paper and apple all rot easily.
- The minerals and compounds return to the soil. They are then used by plants to grow.

Differentiation

- ⇑ 'What organisms help things to rot? What conditions are best for this?'
- ⇓ 'Choose two things from the list that rot and draw them.' You could get the students to draw things in the classroom that would rot in five years. 'Where does all the "stuff" go? Does it disappear?'

Dr McMahon's ice

YEAR 7

Objective covered

QCA 7G. ***Framework P1.***

Aim

To practise drawing the molecular structures of solids, liquids and gases, and to recognise the effect of heat on those structures.

Resources

OHP and OHT, board or worksheet prepared with the following:

Dr McMahon is in the Arctic looking at the structure of water molecules in ice.

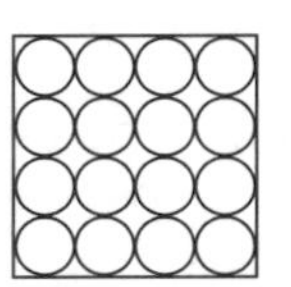

1 What will happen to the solid ice if put in a beaker and heated over a gentle Bunsen flame?
2 What will happen to the structure of the molecules after the ice has been gently heated? Draw them.
3 A strong Bunsen flame is then used. What will eventually happen to the ice which was put in the beaker?
4 What will happen to the structure of the molecules after being strongly heated? Draw them.

Activity

- The students should answer the questions and draw the structure of molecules in water and gas.

Answers

1 Melt

2 

3 Evaporate

4 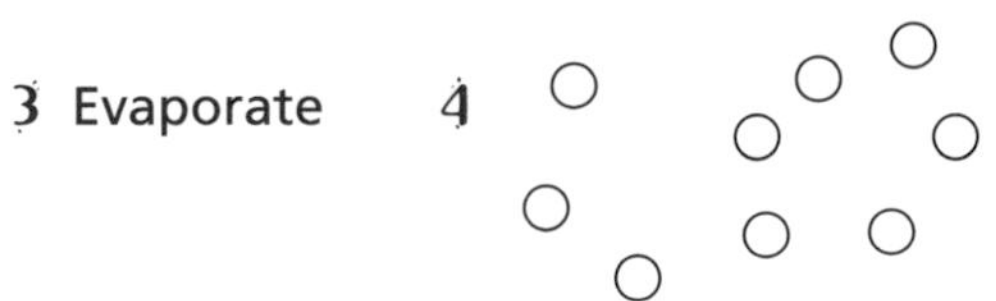

Differentiation

- ⇑ Ask the students to explain why heating a solid will turn it into a liquid and then to a gas.

Roll 'em, roll 'em, roll 'em

YEAR 7

Objective covered

QCA 7G. ***Framework P1.***

Aim

To describe the arrangement of particles in a solid, liquid and gas.

Resources

OHP and OHT, board or worksheet to show the arrangements below with the associated writing frames.

1  2   3 

Complete these words to describe each pattern:

1 R_g_l_ _, f_x_d, o_d_r_d, at_ _ac_ed. This is a s_ _ _ _.
2 R_ _d_m, m_v_n_, less at_ _ac_ed. This is a l_q_ _d.
3 Fast m_v_ _g, n_t t_uch_ _ _, n_ at_ _ _ _t_o_. This is a g_ _.

Activity

- Students should complete the three sentences.

Answers

1 Regular, fixed, ordered, attracted. This is a solid.
2 Random, moving, less attracted. This is a liquid.
3 Fast moving, not touching, no attraction. This is a gas.

Differentiation

- ⇑ Ask students to extend the activity by adding relative kinetic energies of the solid, liquid and gas.
- ⇓ Give the students a prepared worksheet with the completed words to select their answers from.

Spread it about a bit

Objective covered/aim

QCA 7G. ***Framework P2.***

Resources

OHP and OHT, board or worksheet prepared with these two lists:

List A

- Steaming saucepan on a camping stove
- Someone smelling a perfume bottle
- Someone putting sugar in their tea

List B

- Diffusion
- Dissolving
- Boiling

Activity

- Ask the students to draw each item in List A. They should then write the word from List B that describes how the pattern of particles changes in each picture.
- Ask the students to match each of these descriptions to items from Lists A and B. They should then fill in the gaps.

1. Particles of a solid spread throughout a liquid. This is _______________.
2. Particles have sufficient energy to change from a liquid to a ___.
3. Particles move from where there are lots of them to where there are a few of them until they are all evenly s_________ o_____.

Answers

1 Dissolving: sugar in teacup (missing word is 'dissolving')
2 Boiling: saucepan on camping stove (missing word is 'gas')
3 Diffusion: perfume (missing words are 'spread out')

Differentiation

- ⇑ Ask the students to think about energy and rewrite 1, 2 and 3 in terms of energy.

Move it!

Objective covered

QCA 7G. *Framework P2.*

Aim

To recall ideas of patterns of particles to see why gases are compressible, why heating causes expansion and why melting occurs.

Resources

OHP and OHT, board or worksheet prepared with the following:

1 Write a sentence to describe how a gas can change into a solid.
2 Draw the structure of the molecules in a liquid.
3 Write a sentence to describe how a solid changes into a liquid.

Activity

- Ask the students to answer the three questions.

Answers

1 Sentences could include the following words: squash, squeeze, push together, neat pattern and compress.
2 For example:
3 Sentences could include the following words: vibrate, wobble, push apart, expand, break, irregular pattern and melt.

Differentiation

- ⇑ Ask the students to construct a sentence for each of the three questions but to add in the energy involved in each scenario.
- ⇓ Give the students some key words to use in their sentences.

Crystal clear

YEAR 8

Objective covered

QCA 8G. ***Framework P1.***

Aim

To recall that large crystals form from molten materials and solutions on slow cooling.

Resources

OHP and OHT, board or worksheet prepared with the following:

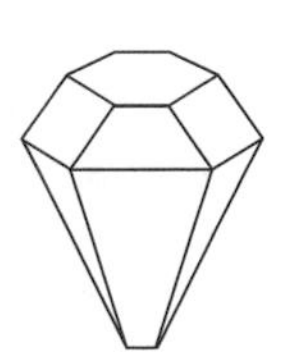

a Sketch of a small crystal.
Write '5 minutes' next to the sketch.

b Sketch of a large crystal.
Write '1000 years' next to the sketch.

Activity

- Ask the students these questions based on the two drawings:
  1. What is the time difference between a and b?
  2. What are the visual differences between a and b?
  3. Write a descriptive sentence to describe why crystals are different sizes.
  4. If you had an evaporating basin half-filled with blue copper sulphate, how would you get large crystals?

Answers

1 1000 years
2 Size and shape of remaining rock
3 Different rates of cooling
4 Very slow cooling

Differentiation

- ⇑ 'Volcanoes produce molten rocks called lava. If a rock has small crystals, where was it likely to have formed? If the rock has large crystals, where is it likely to have formed?'
- ⇓ Ask the students to complete these simple sentences:
  Large crystals form s_ _ _ _ _. Small crystals form q_ _ _ _ _ _.

At the turnstile

Objective covered

QCA 8A. ***Framework P1.***

Aim

To recall that plant and animal cells have a membrane which lets different sized substances in and out of the cell.

Resources

OHP and OHT, board or worksheet prepared with the following:

1 Complete the labels:

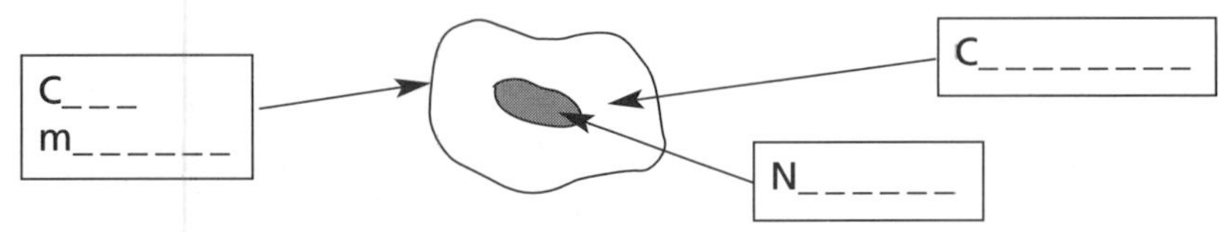

2 Match the names to the functions:

- Nucleus • makes and stores chemicals
- Cytoplasm • holds the cell together
- Cell membrane • controls the cell

3 Some of the chemicals that move in and out of the cell are o_ _ _ _ _, g_ _ _ _ _ _ and c_ _ _ _ _ d_ _ _ _ _ _.

4 Do the chemicals in 3 have all the same-sized particles?

Activity

- Get the students to use their knowledge to answer the questions.

Answers

1 Cell membrane, cytoplasm, nucleus
2 Nucleus – controls the cell
Cytoplasm – makes and stores chemicals
Cell membrane – holds the cell together
3 Oxygen, glucose, carbon dioxide
4 No

I'm special

Objective covered

QCA 8E. ***Framework P2.***

Aim

To develop the idea that there are only a certain number of different types of atom and that these can be found as solids, liquids and gases.

Activity

- Ask the students to count the number of students in the class.
- Then ask them to calculate the total number of Year 8 students in the school by multiplying the number of Year 8 classes by the number of students in each class.
- 'What is the basic unit of Year 8?'
- 'In the UK there are thousands of Year 8 students with many differences between them. However, there are certain things which make all Year 8s different from the rest of the population of the UK. How many differences can you suggest?'

Answers

- The basic unit of Year 8 is a student.
- Differences may include height, weight and gender, but it is the fact that most are the same age that the activity is trying to secure. As with atoms, many humans are similar but there are always going to be unique features.

Differentiation

- ⇑ More able students should be able to give a list of the differences and similarities. Use judgement to conclude the activity and develop the science.
- ⇓ Less able students may only be able to do the counting and multiplication. They may need your help to establish the idea of similar age.

Building blocks

YEAR 8

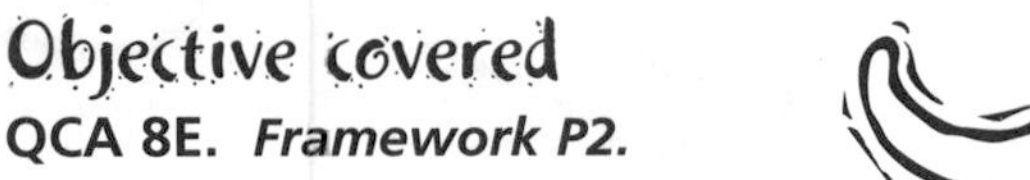

Objective covered

QCA 8E. *Framework P2.*

Aim

To recall that the atom is the basic building block of matter.

Resources

Small squares of white paper, scissors, dark strips of paper and sticky tape.

Activity

- Ask the students to get into pairs. Tell them to cut up the white square of paper into the smallest pieces possible. Give them five minutes to do this.
- After five minutes get them to use a strip of sticky tape to pick up the tiny pieces and fix them onto the dark strip of paper.
- Ask the students these questions:
  1. Are there smaller pieces that the white paper could be cut into? Explain.
  2. What do you think the smallest piece is made from?

Answers

1 Yes – students should recall that invisible atoms build up visible matter.
2 Atoms

Differentiation

- ⇑ Ask the students to identify the building block atoms in glucose $C_6H_{12}O_6$, water H_2O and carbon dioxide CO_2.
- ⇓ Ask the students to use the building blocks C, O and H to make CO_2 and H_2O.

My one and only

YEAR 8

Objective covered

QCA 8E. ***Framework P2.***

Aim

To recall that elements consist of only one type of atom.

Resources

OHP and OHT, board or worksheet prepared with the following sentences:

We breathe in o_ _ _ _ _ and breathe out c_ _ _ _ _ d_ _ _ _ _ _.
We use salt on food. The chemical name for sea salt is s_di_m chlor_de.

Activity

- Ask the students to complete the sentences.
- Next, ask these questions about the completed sentences:
  1. Which of the words you completed have only one type of atom?
  2. Which of the words you completed have more than one type of atom?
  3. Which of the words you completed are elements?

Answers

The missing words are 'oxygen', 'carbon dioxide' and 'sodium chloride'.

1 Oxygen
2 Carbon dioxide and sodium chloride
3 Oxygen

Differentiation

- ⇑ Ask the students to list ten everyday objects in the kitchen and garden and decide if they are made from elements.
- ⇓ Provide these sentences: Saucepans can be made from i_ _n, alu_i_iu_ or co_ _e_. These are el_ _ _ _ _ _ because they are made of only one t_ _e of at_ _. (iron, aluminium, copper, elements, type, atom)

Some for all

Objective covered/aim

QCA 8E. *Framework P2.*

Resources

OHP and OHT, board or worksheet prepared with the following.

Some elements in everyday life are:

- Hydrogen and oxygen – in water – H_2O
- Carbon and oxygen – in carbon dioxide – CO_2
- Carbon and hydrogen – in natural gas called methane – CH_4
- Carbon and oxygen – in carbon monoxide – CO
- Nitrogen and hydrogen – in ammonia – NH_3

Activity

- Ask the students to answer these questions:
  1. How many of the compounds contain hydrogen?
  2. How many of the compounds contain oxygen?
  3. How many of the compounds contain carbon?
  4. How many different elements are the five compounds made from?

Answers

1 3   2 3   3 3   4 5

Differentiation

- ⇑ Ask the students to use the numbers 3, 4 and 5 to make as many sums as possible. They should see that there are many combinations possible from a few starting items.
- ⇓ Ask the students to memorise names of the elements, compounds and formulae by the method of 'look, cover, write, check'. They should test their neighbours.

Elemental my dear Watson

YEAR 8

Objective covered

QCA 8E. ***Framework P2.***

Aim

To be able to recognise the difference in structure between an element, mixture and a compound.

Activity

- Use coloured spots, drawn on the board, to represent atoms in elements, mixtures and compounds.
  For example:

  1   2   3   4

- Ask the students to identify the diagrams as elements, mixtures or compounds.

Answers

1 Element   2 Compound   3 Mixture   4 Mixture

Differentiation

- ⇑ Give the students diagrams with three different atoms.
- ⇑ Draw a mixture of compounds, for example:

Stinky eggs

Objective covered/aim

QCA 8F. ***Framework P3.***

Resources

OHP and OHT, board or worksheet prepared with the following:

iron + sulphur → iron sulphide

1 Name the two elements.
2 Name the compound.
3 Identify a difference between the answers in 2 and 3.
4 What is the ratio of iron to sulphur in iron sulphide?
5 What does iron sulphide smell like?

Activity

- Get the students to use their knowledge to answer the questions.

Answers

1 Iron and sulphur
2 Iron sulphide
3 Iron + sulphur is magnetic but iron sulphide loses magnetism through heating.
4 1 : 1
5 Rotting eggs

Differentiation

- ⇑ 'If one iron atom combines with one sulphur in the ratio of 1 : 1, what will be the ratio of hydrogen to oxygen in water, carbon to oxygen in carbon dioxide, and copper to sulphate in copper sulphate?'
- ⇓ Ask the students to form the word equation for hydrogen and oxygen making water. They could draw pictures of hydrogen and oxygen and then water. Use of a textbook or sheets that have the information may help.

All change please

Objective covered/aim

QCA 9F. ***Framework P1.***

Resources

OHP and OHT, board or worksheet prepared with the following:

wood + oxygen → carbon dioxide + water + HEAT

1. After burning, describe how the wood has changed.
2. Can we reverse the process and un-burn the wood? Explain your answer.
3. Does the wood have a different number and arrangement of atoms compared to carbon dioxide and water?
4. Are new chemicals made? If so, how many? If not, why not?
5. Has energy been transferred? How?
6. Complete the following:
   When a _____ burns, a ___________ change takes place and _______ is transferred.

Activity

- Ask the students to answer the questions based on the equation.

Answers

1. Any suitable description
2. No, this is a non-reversible reaction.
3. Yes
4. Yes, 2 (carbon dioxide and water)
5. Yes, chemical potential energy → heat and light
6. Missing words are 'fuel', 'chemical' and 'energy'.

Differentiation

- ⇓ Get the students to draw and label what is left when wood burns, coal burns, oil burns and when gas burns. They should try to see differences and similarities between each case. Encourage them to label the fact that energy has been transferred.

Check your change

Objective covered/aim

QCA 9F. ***Framework P2.***

Activity

- Write this equation on the board:

  magnesium + hydrochloric acid → magnesium chloride + hydrogen

- Ask the students the following questions:

  1 Has there been a change during the reaction?

  2 How would you confirm that hydrogen gas is being given off?

  3 Have the atoms rearranged themselves during the reaction?

Answers

1 Yes, there has been a change. A solid has dissolved, a gas has been given off, there would be a smell.
2 'Squeaky pop' test for hydrogen.
3 Yes. The hydrogen has rearranged itself from a liquid form to a gas form. The magnesium has changed from a solid to a liquid solution where it arranged itself with the chloride from the acid.

Differentiation

- ⇑ Give the students the equation that shows how acid rain attacks limestone buildings:

  acid rain + limestone → calcium sulphate + water + carbon dioxide
  H_2SO_4 + $CaCO_3$ → $CaSO_4$ + H_2O + CO_2

  Ask them what has been rearranged and where.

- ⇓ Ask the students to draw pictures of acid rain damaging buildings. They should be able to label that the limestone dissolves with the acid rain and that carbon dioxide forms from the limestone.

I bet your name is...

Objective covered/aim

QCA 9E. ***Framework P3.***

Resources

OHP and OHT, board or worksheet prepared with the following:

1 Link up these names and formulae:

Carbon dioxide	MgO
Water	$CuCO_3$
Magnesium oxide	CO_2
Copper carbonate	CH_4
Natural gas	H_2O

2 Using these symbols, O_2, Fe, H_2, H_2O, FeS, Mg, MgO, S, write symbol equations for the following reactions:

a iron + sulphur → iron sulphide

b hydrogen + oxygen → water

c magnesium + oxygen → magnesium oxide

Activity

- Tell the students to get into pairs.
- Ask them to work together to answer the questions.

Answers

1 Carbon dioxide, CO_2; water, H_2O; magnesium oxide, MgO; copper carbonate, $CuCO_3$; natural gas, CH_4

2a $Fe + S \rightarrow FeS$ b $2H_2 + O_2 \rightarrow 2H_2O$ c $2Mg + O_2 \rightarrow 2MgO$

Differentiation

- ⇑ Give the formulae for sodium hydroxide, hydrochloric acid, sodium chloride and water and ask the students to construct the symbol equation for the reaction of sodium hydroxide and hydrochloric acid.
- ⇓ Ask the students to predict the name of the compound made when zinc and oxygen react.

The rules are...

YEAR 9

Objective covered/aim

QCA 9E. Framework P3.

Resources

OHP and OHT, board or worksheet prepared with a picture of a beaker half-filled with liquid labelled 'copper sulphate' and a picture of a spatula carrying a small heap labelled 'zinc'.

Activity

- Ask the students to use their knowledge to answer these questions:

  1. What happens when zinc is added to copper sulphate? (Remind them of the word 'displacement'.)
  2. Is any mass lost when the zinc reacts with the copper sulphate?
  3. Has a chemical reaction occurred between zinc and copper sulphate?
  4. If the beaker, copper sulphate and zinc together have a mass of 70 g at the beginning of the reaction, what will be the mass at the end of the reaction?
  5. Write a word equation to show the reaction.
  6. Is the mass balanced in a chemical reaction like the one above?

Answers

1. Zinc displaces the copper from the solution.
2. No
3. Yes, heat is given out and the colour changes from blue to colourless with red/brown copper forming.
4. 70 g
5. Zinc + copper sulphate → zinc sulphate + copper
6. Yes

Big bullies

YEAR 9

Objective covered

QCA 9F. *Framework P4.*

Aim

To practise writing displacement reactions.

Activity

- Write on the board suitable word equations for metal reactions with salts, for example:

lithium + iron oxide → __________ + __________ __________

sodium + copper oxide → __________ + __________ __________

calcium + lead oxide → __________ + __________ __________

- Ask the students to fill in the missing words.
- Alternatively, change the position of the missing words, for example:

__________ + magnesium oxide → __________ + sodium oxide

Answers

lithium + iron oxide → iron + lithium oxide
sodium + copper oxide → copper + sodium oxide
calcium + lead oxide → lead + calcium oxide
sodium + magnesium oxide → magnesium + sodium oxide

Differentiation

- ⇑ The students could try to write the equations using chemical symbols.
- ⇓ Ask the students to draw a quick cartoon to show the 'stronger' element bullying the 'weaker' element from its partner.

Metal reactions - oxygen

YEAR 9

Objective covered

QCA 9E. ***Framework P4.***

Aim

To practise simple word equations of metal reactions with oxygen.

Activity

- Write on the board word equations of metal reactions with oxygen, for example:

  lithium + oxygen → __________ __________

  sodium + oxygen → __________ __________

- Ask the students to fill in the missing words.
- Alternatively, change the position of the missing words, for example:

  __________ + oxygen → sodium oxide

Answers

lithium + oxygen → lithium oxide
sodium + oxygen → sodium oxide

Differentiation

- ⇑ The students could try to write the equations using chemical symbols.
- ⇓ 'How do you test for oxygen?'

Metal reactions - water

YEAR 9

Objective covered

QCA 9E. ***Framework P4.***

Aim

To practise simple word equations of metal reactions with water.

Activity

- Write on the board word equations of metal reactions with water, for example:

  lithium + water → __________ __________ + __________

  sodium + water → __________ __________ + __________

- Ask the students to fill in the missing words.
- Alternatively, change the position of the missing words, for example:

  __________ + water → sodium hydroxide + __________

Answers

lithium + water → lithium hydroxide + hydrogen
sodium + water → sodium hydroxide + hydrogen

Differentiation

- ⇑ The students could try to write the equations using chemical symbols. For example, write:
  2__________ + $2H_2O$ → 2NaOH + 2__________
- ⇓ 'How do you test for hydrogen?'

Metal reactions - acid

YEAR 9

Objective covered

QCA 9E. ***Framework P4.***

Aim

To practise working out which salt is produced when a metal reacts with an acid.

Resources

OHP and OHT prepared with a grid with the names of various metals and acids and their associated salts, for example:

	Magnesium	**Iron**	**Sodium**	**Lithium**
Hydrochloric acid	Magnesium chloride	Iron chloride	Sodium chloride	Lithium chloride
Sulphuric acid	Magnesium sulphate	Iron sulphate	Sodium sulphate	Lithium sulphate
Nitric acid	Magnesium nitrate	Iron nitrate	Sodium nitrate	Lithium nitrate

Cover the names of the salts with individual pieces of paper.

Activity

- Ask the students (individuals or teams) to call out the names of an acid and metal combination. They then tell you the name of the salt that is produced.
- The piece of paper covering the combination is then removed to reveal the answer. If they get it right, a point is awarded (if not the point goes to the other team/individuals).

Differentiation

- ⇑ Use chemical symbols instead of words.

Reactive metals

Objective covered/aim

QCA 9F. ***Framework P5.***

Resources

OHP and OHT prepared with the following grid:

	Reaction with oxygen	Reaction with water	Reaction with acid
Iron	Slow reaction	Very slow reaction	Reacts
Magnesium	Burns very quickly	Reacts only with steam	Reacts
Gold	No reaction	No reaction	No reaction
Copper	Slow reaction	No reaction	No reaction
Sodium	Burns	Reacts rapidly with cold water	Reacts violently

Activity

- Ask the students to work out the order of reactivity for these metals using the information in the grid.

Answers

Sodium — Most reactive
Magnesium
Iron ↓
Copper
Gold — Least reactive

Differentiation

- ⇑ Get the students to suggest uses for the metals, for example iron is used in construction.
- ⇓ Tell the students to memorise the five metals in this reactivity series.

Open before use

YEAR 7

Objective covered

QCA 7K. ***Framework F1.***

Aim

To recall that forces act in directions and can be big or small.

Resources

OHP and OHT, board or worksheet prepared with a picture of a tube of toothpaste, tin of paint and a jam jar.

Activity

- Ask the students which of the containers is easiest/hardest to open.
- Get them to draw diagrams with arrows to show which way the forces act when opening each container.
- The students should write sentences using the phrases 'size of force' and 'direction of force' to go with their diagrams.

Answers

The toothpaste tube is the easiest to open and arrows should be rounded.
The jam jar is next and again arrows should be rounded.
The tin of paint is hardest to open and arrows should be upwards for the lid and downwards for the lever to lift the lid.

Differentiation

- ⇑ 'If 5 N is needed to tear a tissue, estimate the force needed to open the toothpaste, the paint and the jam jar.'
- ⇓ Tell the students to use the words 'big', 'bigger' and 'biggest' to label the forces needed to open the toothpaste, tin of paint and jam jar.

Keep the right balance

Objective covered

QCA 7K. *Framework F2.*

Aim

To recall that when forces are not equal things change direction.

Activity

- Ask the students whether the forces in these are balanced or unbalanced:

  1 A book on a table

  2 A pool ball colliding with another pool ball

  3 A car moving at constant speed along a straight road

  4 A skidding car

Answers

1 Balanced
2 Unbalanced
3 Balanced
4 Unbalanced

Differentiation

- ⇑ Ask the students to identify the forces acting using as many words as they know about forces.
- ⇓ Give the students a worksheet with pictures of 1 to 4 and ask them to label the pictures with the forces.

Unbalanced forces

YEAR 7

Objective covered

QCA 7K. ***Framework F2.***

Aim

To recall that unbalanced forces lead to a change in speed.

Resources

OHP and OHT, or board prepared with the following diagrams:

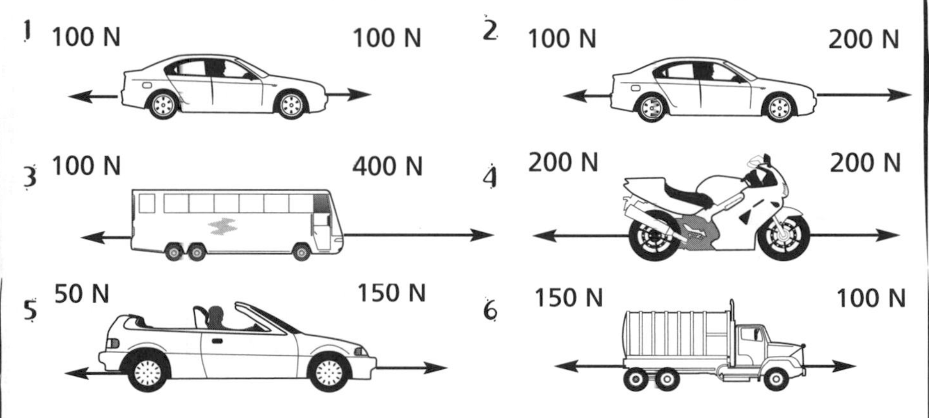

Activity

- Ask the students to decide whether the vehicles are accelerating, moving at a steady speed or decelerating.

Answers

1 Steady speed 2 Accelerating 3 Accelerating
4 Steady speed 5 Accelerating 6 Decelerating

Differentiation

- ⇑ Ask the students to identify which of Newton's laws the diagrams illustrate.
- ⇓ Give the students the diagrams on a worksheet.

Nobody move!

Objective covered

QCA 7K. *Framework F3.*

Aim

To practise recognising the forces acting on a stationary object.

Resources

A 'large interesting object' (like a vase or a toy gorilla) positioned on your desk.

Activity

- Ask the students to:
  - draw the 'large interesting object'
  - draw an arrow to show the force that stops the 'large interesting object' from floating off
  - draw an arrow to show the force that stops the 'large interesting object' from falling to the ground.

Answers

The sketch should have gravity acting downwards and upthrust from the table acting upwards. The arrows should be of equal length to show equal force.

Differentiation

- ⇑ Ask the students to explain why the object will begin to fall if the table is removed.
- ⇓ Provide the students with a pre-drawn diagram on a worksheet and/or cut and paste labelled arrows, for example:

↓ Force of gravity stops vase from floating off table.

How fast?

Objective covered

QCA 7K. ***Framework F4.***

Aim

To practise using the equation for working out speed.

Activity

- Ask the students these questions:

1. A car takes 2 hours to drive from Bristol to London. The distance is 100 km. At what speed was the car travelling?
2. I can run 100 m in 10 seconds. At what speed can I run?
3. Superman flies 3000 km in 2 hours. At what speed was Superman flying?
4. My Gran can run the 200 m race in 40 seconds. At what speed can she run?

Answers

1 50 km/h   2 10 m/s   3 1500 km/h   4 5 m/s

Differentiation

- ⇑ Provide the students with the equation triangle and mix up the questions to work out distance and time, for example:
  I am driving at 50 km/h. The journey is 100 km. How long will it take me to finish the journey?

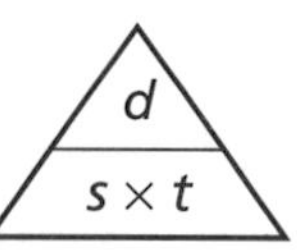

- ⇓ Provide the questions on a worksheet.
- ⇓ Provide a word equation, i.e. speed = distance ÷ time.

Mass watchers

Objective covered/aim

QCA 7L. ***Framework F5.***

Resources

OHP and OHT, board or worksheet prepared with the following text and questions:

> Hello, my name is Zog, and I'm from the planet Pluto. I am visiting planet Earth and I have a problem. I am made of 70 kilograms of matter, I call this my *mass*. My new friends on Earth asked 'What *weight* are you?' To this I replied 'Seven hundred newtons', and they all laughed. Please help me tell them, in your language, what the difference between *mass* and *weight* is.

1. Write a sentence to explain the difference between weight and mass.
2. What is the force of gravity on Earth?
3. On Pluto, Zog weighs 40.6 N. What is the force of gravity on his home planet?

Activity

- Ask the students to use their knowledge to answer the questions.

Answers

1. Mass is the amount of matter in an object (measured in kg). Weight is a force caused by the pull of gravity on a mass (measured in N).
2. Force of gravity on Earth is 10 N/kg.
3. Force of gravity on Pluto is 0.58 N/kg.

Differentiation

- ⇑ Give the gravity force on the Moon (1.6 N/kg) and the weight of 112 N and ask the students to find the mass (70 kg).
- ⇓ The students could construct a sentence using the following key words: 'difference', 'times', 'bigger', 'mass', 'between', 'weight'.

The Professor's friction

YEAR 7

Objective covered

QCA 7K. ***Framework F6.***

Aim

To practise identifying friction in everyday scenarios.

Resources

OHP and OHT, board or worksheet prepared with the following letter:

> *Dear students*
>
> *My name is Professor Banks. I am designing the world's fastest car. I have recently heard of a phenomenon called 'friction'. Please help me by answering the following questions.*
>
> *1 Draw a racing car. (2 minutes)*
>
> *2 Label the parts of the car where friction is found.*
>
> *3 Tell me how friction can be reduced in these areas.*
>
> *Thank you for your help.*
>
> *Professor Banks*

Activity

- Ask the students to answer the questions in Professor Banks' letter. You may wish students to answer the questions as a reply letter to Professor Banks.

Differentiation

- ⇑ Ask the students where friction is needed in a car (brakes, between the tyres and the road and so on).
- ⇓ Provide the students with labels to put on their diagram, for example:

Friction in ball bearings	Friction reduced by using oil

It's magic

YEAR 8

Objective covered

QCA 8J. ***Framework F1.***

Aim

To recall that magnetic materials exert a force of attraction and repulsion.

Resources

OHP and OHT, board or worksheet prepared with the following:

1 What happens when two magnets are arranged as follows?

S	N	N	S		N	S	N	S

2 Use the following words to make a sentence to describe what happens.

attract	repel	touch	pole
move apart	don't touch	move together	

3 What things do you know that are magnetic?

Activity

- Ask the students to use their knowledge to answer the questions.

Answers

1 Two north poles (or two south poles) will repel. One north and one south pole will attract.
2 Answer by outcome.
3 Answers by outcome. Metals must contain iron or steel.

Differentiation

- ⇑ 'Do horseshoe magnets work in the same way as bar magnets? Explain your answer.'
- ⇓ Work with less able students to construct the sentence and then play a memory game in pairs – practise saying the sentence and see which pair is the first to repeat their sentence precisely.

Always, sometimes or never

YEAR 8

Objective covered

QCA 8J. ***Framework F1.***

Aim

To practise recognising magnetic materials.

Resources

OHP and OHT, board or worksheet prepared with pictures of:

1 a bar magnet
2 an iron nail
3 a bicycle labelled 'aluminium'
4 a nail with copper wire wrapped around it, linked to a battery (electromagnet)
5 an aluminium drinks can.

Activity

- Ask students to label the pictures with 'always', 'sometimes' or 'never' for the pull or push that they feel when a magnet is put close to the objects.

Answers

1 Always   2 Always   3 Never   4 Sometimes   5 Never

Differentiation

- ⇑ Develop 4 by asking the students to show how to wire up the circuit and explain the idea of an electromagnet.
- ⇓ Ask the students to sketch bar magnets with the magnetic field lines to show attraction and repulsion:

N	S		N	S	and	N	S		S	N

That's nailed 'em

Objective covered

QCA 8J. ***Framework F3.***

Aim

To recall how magnetic field patterns change when the strength of an electromagnet increases.

Resources

OHP and OHT, board or worksheet prepared with illustrations of:

a a nail with 4 turns of wire connected to a battery labelled 6 V, for example:

6 V

b a nail with 2 turns of wire connected to a battery labelled 6 V
c a nail with 8 turns of wire connected to a battery labelled 6 V
d a nail with 12 turns of wire connected to a battery labelled 6 V.

Activity

- Ask the students these questions about the diagrams. They should explain their answers:
  1 Which nail has the closest magnetic field lines?
  2 Which nail has the greatest distance between the magnetic field lines?

Answers

1 The closest magnetic field lines are found in d because this has the largest number of turns.
2 The greatest distance between the magnetic field lines is found in b because this has the lowest number of turns.

Differentiation

- ⇑ The students could discuss the 'fair test' concept in this activity.
- ⇓ Prepare this sentence for the students to complete, for example:
  Nail ____ has the closest magnetic field lines because it has ____________ coils around it.

Shape up

Objective covered

QCA 9K. ***Framework F1.***

Aim

To recall that friction slows down solids, liquids and gases.

Resources

OHP and OHT, board or worksheet prepared with the following:

1 Rearrange these liquids in order of fastest to slowest-flowing:

golden syrup   engine oil   water   nail-varnish remover

2 Use the following words to construct a sentence explaining the difference between the liquids in question 1.

resistance   density   friction   force   thick   thin   particles

3 Which shape will fall the fastest through water? Explain why.

a   b 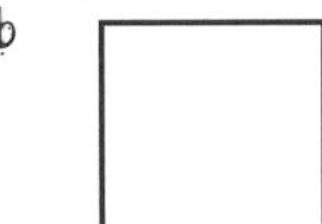   c 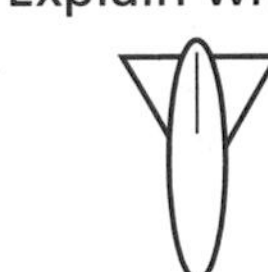

Activity

- Ask the students to answer the questions.

Answers

1 Nail-varnish remover, water, engine oil, golden syrup.
2 The sentence should include friction for slowing liquids down.
3 c is the most hydrodynamic.

Differentiation

- ⇑ Quote values of density for each liquid and ask more able students to relate these densities to their rate of flow.

  Water = 1.0 g/cm^3, engine oil = 1.9 g/cm^3, golden syrup = 2.3 g/cm^3, nail-varnish remover = 0.87 g/cm^3

- ⇓ Ask the students to design a very aerodynamic car.

Just a moment

YEAR 9

Objective covered

QCA 9L. ***Framework F2.***

Aim

'To see a see-saw as much more than you first saw.
To use gravity and forces, distances and pivots and
see in a moment how your physics will soar!'

Activity

- Write these anagrams on the board:

  gdra tthsru crtiinof ytivrga

- Ask the students to unscramble the words.
- Get the students to sketch a see-saw and show where they would sit if they had a friend who was much shorter and lighter than them sitting on the other end of the see-saw.
- Ask the students to suggest where a heavier senior citizen should sit on the see-saw. They should explain their answer.

Answers

- drag, thrust, friction, gravity
- Sketches should have the lighter student further away from the centre (students should use the word pivot or fulcrum).
- With the heavier senior citizen on the see-saw the distances from the pivot should change. More able students should consider the combined force due to both the friends sitting on one side of the pivot as well as the individual forces due to each young person sitting individually opposite the senior citizen.

Differentiation

- ⇑ The students could use specimen masses of 50 kg for themselves, 35 kg for the lighter friend and the senior citizen at 80 kg. They must convert these to forces, i.e. 500 N, 350 N and 800 N.
- ⇓ The students could simply label the see-saw with 'up' and 'down' and the relative positions where each person would sit.

Get the point

YEAR 9

Objective covered/aim

QCA 9L. *Framework F3.*

Resources

OHP and OHT, board or worksheet prepared with the following:

'I helped build the path up the garden. I noticed that the flat slabs didn't sink as far as those that were propped up. I think I understand why, but could you help me choose some of the following words to write a sentence explaining why?'

mass	length
weight	width
force	area
side	pressure

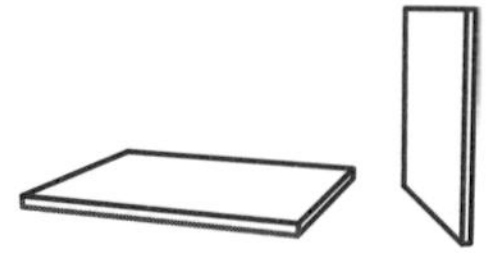

Activity

- Get the students to write a sentence responding to the problem.

Answers

The greatest pressure is measured under the slab on its side as this has the least area through which the downward force (weight) is acting.

Differentiation

- ⇑ Ask the students to answer this question:

  The slab measures 50 cm by 20 cm by 3 cm. The mass of the slab is 20 kg. Find the pressure underneath:

  1 the flat slab (0.2 N/cm^2)
  2 the slab resting on its longest edge (1.3 N/cm^2).

- ⇓ The students could fill in the sentence below:

  The propped up paving slab has a s_ _ _ _ _ _ surface area through which the f_ _ _ _ is acting and so exerts a greater p_ _ _ _ _ _ _ and sinks further into the ground.

  (smaller, force, pressure)

Massive weights

YEAR 9

Objective covered

QCA 7K. Framework F4.

Aim

To practise working out the weight of an object when its mass and the force of gravity are known.

Resources

Bathroom scales measuring in kilograms.

Activity

- Display information about the force of gravity on different planets:

  Mercury = 4 N/kg Mars = 4 N/kg Venus = 9 N/kg
  Jupiter = 26 N/kg Earth = 10 N/kg Saturn = 11 N/kg
  (Moon = 1.6 N/kg, but not a planet)

- You may also wish to include the equation:

  Weight (N) = mass (kg) × force of gravity (N/kg)

- Choose one of these activities:
  - Provide the students with a list of masses and ask them to work out the weight of those masses on the different planets.
  - Ask the students to measure their masses in kilograms and then work out their weights on the different planets.

Differentiation

- ⇑ Provide the mass of an object and its weight on a planet and ask the students to work out the force of gravity on that planet.
- ⇑ A formula triangle could be provided:

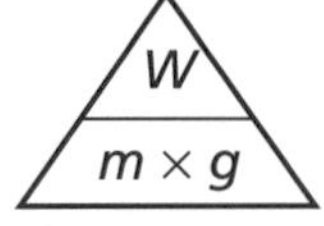

- ⇓ The students could be provided with questions and equations on a worksheet. For example, _____ N = 100 kg × 10 N/kg.

Size does matter

Objective covered

QCA 9J. ***Framework F4.***

Aim

To recall and establish that *size* (mass) affects *gravity* which affects *weight*. The bigger the object the greater the force of gravity.

Resources

OHP and OHT, board or worksheet prepared with the following:

1 NASA are designing two rockets. Which will have the greatest weight whilst on planet Earth, the large or the small one?
2 Explain your answer to question 1 using the word 'gravity'.
3 As the astronauts travel away from planet Earth they become lighter until they become weightless. What does this tell you about the effect of the Earth's gravity as you get further from the planet?
4 There are two astronauts onboard the rocket, a big astronaut and a small astronaut. Which will have the greatest weight whilst in outer space?

Activity

- Ask the students to answer the questions.

Answers

1 The large rocket will have the greatest weight.
2 The large rocket has the greater mass therefore experiences the greater force due to gravity and therefore has the greater weight.
3 The further from the planet the less the force of gravity.
4 They will both be weightless.

Differentiation

- ⇑ 'What other gravitational forces in the solar system will affect an astronaut?' (Moon, Sun and other planets.)
- ⇓ Give the students a writing frame for the answers, for example:

  The l_ _ _ _ _ _ rocket will have the greatest weight.

Fuels forever

Objective covered

QCA 7I. Framework E1.

Aim

To recall the fuels that we use daily.

Resources

OHP and OHT, board or worksheet prepared with the following questions:

1 Make a list of as many fuels as you can.
2 Underline the fossil fuels.
3 Put a star by the fuels that you use every day.
4 Fossil fuels will not last forever. If you were the Prime Minister, how would you encourage the people of this country to use less fossil fuels?

Activity

- Ask the students to answer the questions.
- You may wish the students to carry out this activity in small groups.

Differentiation

- ⇑ The students could prepare a presentation, new set of laws, or a letter to the Prime Minister on the alternative energy resources.
- ⇓ Ask the students to draw fuels in use.

Alien energy

Objective covered/aim

QCA 7I. *Framework E2.*

Resources

OHP and OHT, board or worksheet prepared with a letter from an alien.

Dear Earthlings

My name in Zoga, I am the president of the planet Zargon. On my planet everything is dark. We have wires, but nothing flows through them. Nothing ever falls down. Everything is motionless. If you shout, no one will hear you. Everything is frozen at 'absolute zero' (–273°C). Elastic bands will not fire if pulled and rubber balls do not bounce. Finally, we have no chemical fuels like petrol or food like apples.

I have read that on planet Earth you have something called energy and that all this energy originates from the Sun. Please can you write to me to tell me the types of energy you have on planet Earth.

All the best
Zoga

Activity

- Ask the students to write a postcard to reply to Zoga's letter telling her about the eight types of energy on planet Earth.

Answers

The eight types of energy are light, electrical, gravitational, kinetic, sound, thermal, potential and chemical.

Differentiation

- ⇑ More able students could write a letter.
- ⇑ A more able class could be asked to do an imaginative piece of writing, without having seen the letter, to describe a world without energy.
- ⇓ The students could be provided with a sentence, with blanks for the energy types, to be filled in.

Electrical components

Objective covered

QCA 7J. *Framework E2.*

Aim

To practise using the symbols which represent electrical components.

Activity

- Draw a list of electrical component symbols on the board and ask students to identify them.
- Alternatively, you could write a list of electrical components on the board and ask students to draw the symbol for them. For example, bulb, switch, cell, ammeter.
- Ask the students to use the symbols to draw a circuit.

Differentiation

- ⇑ Specify if you want a series or a parallel circuit drawn.
- ⇑ Specify what you want the circuit to do. For example, 'A flight of steps needs to have a switch at the bottom and top of it with a single bulb in the ceiling above the middle of the staircase.'
- ⇓ Prepare the activity on paper for less able students.

Warm nose, warm hands

Objective covered

QCA 8I. ***Framework E1.***

Aim

To recall how particles behave when energy is transferred to them.

Activity

- On the board, draw a narrow-necked round-bottomed flask with the liquid level up the narrow neck, for example:

- Tell the students that when Assa held the flask in his hand the liquid level rose up the narrow neck.
- Ask the students to write an explanation using suitable words from this list:

contracted	warmed	molecules	expanded
energy	thermal	kinetic	

Answers

Thermal energy from Assa's hand is transferred to the water molecules, which move faster as their kinetic energy increases. As the water molecules are moving faster they move further apart and so move up the tube.

Differentiation

- ⇑ More able students must incorporate energy transfers using thermal energy and kinetic energy.
- ⇓ Less able students could use 'warmed-up' and 'expanded' in their sentence. You could give them a simple writing frame to complete.

A little light work

YEAR 8

Objective covered

QCA 8K. ***Framework E2.***

Aim

To recall that we see objects due to reflected light.

Resources

OHP and OHT, board or worksheet prepared with a line drawing of an everyday scene with sources of light and reflectors of light (for example, light bulbs and chairs).

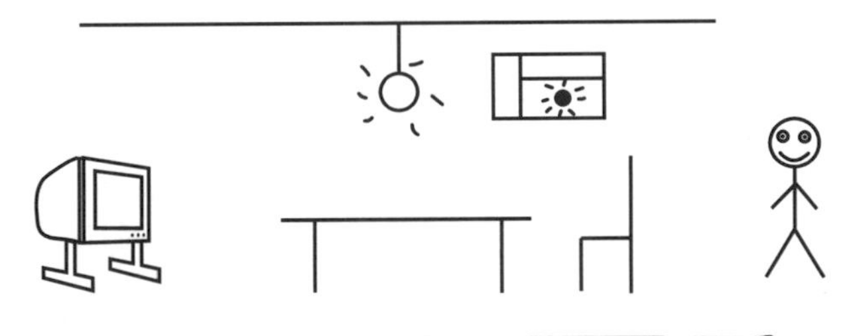

Activity

- Ask the students to draw lines to show how the stick person sees the objects in the room.
- A ruler should be used to draw straight lines from the light source to the light-reflecting object to the stick person's eyes, for example:

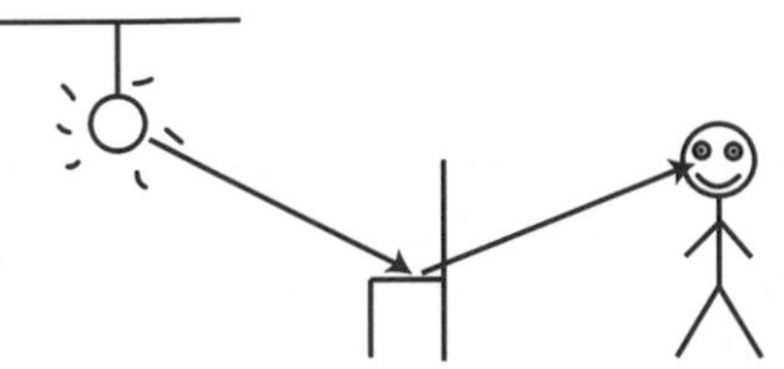

Differentiation

- ⇑ Ask the students to explain why, if light is white, red objects appear red and blue objects appear blue.
- ⇓ Less able students will need to be provided with the picture on a worksheet.

Cool vibrations

YEAR 8

Objective covered

QCA 8L. ***Framework E3.***

Aim

To explain the transmission, production and reception of sound.

Resources

OHP and OHT or board prepared with the following diagram:

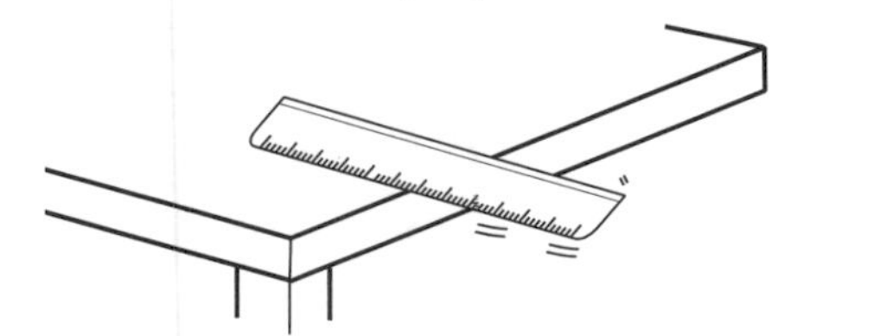

Activity

- Ask the students to copy the diagram into their books.
- Ask them to write a short explanation of how sound reaches the ear from the vibrating ruler.
- Write the statement 'In space no one can hear you scream!' on the board.
- Ask the students to explain why this is the case.

Differentiation

- ⇑ More able students could explain how volume and pitch are changed. Key words are 'amplitude' and 'frequency'.
- ⇓ Provide less able students with a copy of the diagram on a worksheet.
- ⇓ Provide the students with key words or a writing frame. Key words: 'vibrate', 'air particles', 'energy', 'movement'.

What a state

YEAR 8

Objective covered

QCA 8I. ***Framework E4.***

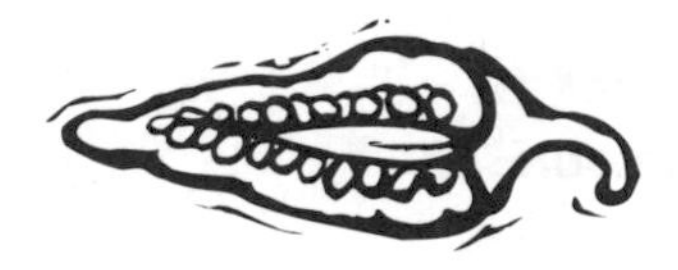

Aim

To practise drawing the structure of molecules in solids, liquids and gases.

Activity

- Ask the students to use nine big dots or circles to draw the structure of molecules in a solid, a liquid and a gas.
- Draw this diagram on the board:

Solid ⇄ Liquid ⇄ Gas

- Remind the students that 'solid', 'liquid' and 'gas' are the three states of matter.
- Say that matter can change state and ask what we call these changes in state.

Answers

Solid  Liquid  Gas 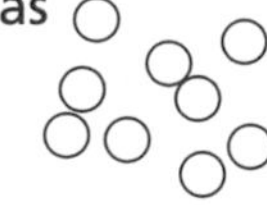

	Melting →		Boiling →	
Solid		Liquid		Gas
	← Freezing		← Condensing	

Differentiation

- ⇑ Ask the students if it is possible to change state from solid directly to gas or from gas directly to solid. (It is – it is called 'sublimation' in both directions.)
- ⇓ Give the students the change of state diagram to write on.

Energetic transfers

YEAR 9

Objective covered

QCA 9I. ***Framework E1.***

Aim

To practise writing energy transfer diagrams.

Resources

OHP and OHT, board or worksheet prepared with a range of everyday scenarios involving energy transfer. Examples: an electric lamp, a moving petrol car, a gas cooker, a falling ball and so on.

Activity

- Ask the students to draw an energy transfer diagram for each of the displayed scenarios.

Differentiation

- ⇑ Ask the students to write energy transfer diagrams for equipment with several stages of energy transfer. For example, for a steam engine:

  Chemical energy ⟶ Heat energy ⟶ Kinetic energy

- ⇑ Get the students to think about energy dissipation in each energy transfer. For example, for an electric light bulb:

  Electricity ⟶ Light
  Electricity ⟶ Heat

- ⇑ More able students could consider energy efficiency. For example:

  Electricity 100 J ⟶ Light 3 J
  Electricity 100 J ⟶ Heat 97 J

- ⇓ Provide the students with a worksheet with parts of simple energy transfers missing. For example, for an electric light bulb:

  Electricity ⟶ __________

You've got potential

YEAR 9

Objective covered/aim

QCA 9I. ***Framework E2.***

Resources

OHP and OHT, board or worksheet prepared with pictures of an iron, a light bulb, a loudspeaker and a DC motor.

Activity

- Display the pictures.
- Ask the students the following questions:

  1. Write down as many types of energy as you can.
  2. Describe the energy transfers that are happening in each of the pictures.
  3. Are these energy transfers 100% efficient?
  4. Is the input and output energy the same for each one? Explain your answer.
  5. What do you now think 'potential difference' means?

Answers

1. The students should include eight types of energy: sound, heat, electrical, light, chemical, potential, kinetic, gravitational.
2. Iron – electrical to heat
   Lamp – electrical to heat and light
   Loudspeaker – electrical to sound
   DC motor – electrical to movement
3. No – no energy transfer is 100% efficient.
4. No – energy transfers always have heat losses and so 100% transfer is not possible.
5. By outcome, but the students should relate potential difference to energy usage by the appliance itself and say that there is a difference in input and output.

Differentiation

- ⇑ Ask the students to estimate percentage transfer of input electrical energy to heat, light, sound and movement with associated percentage energy losses from each appliance.

Just an idea

YEAR 7

Objective covered

Framework SE1.

Aim

To recall that scientific explanations often come from simple ideas or observations.

Resources

OHP and OHT, board or worksheet prepared with the following lists:

Early idea/discovery	**Modern explanation/use**
Beads formed by fire in desert areas	The nucleus
A falling apple	Evolution
The Sun rising and setting	The aeroplane
Bird wing shape	Glass
Pollen grains vibrating on water	The Earth spinning on its axis
Fossils	Brownian motion and diffusion
Dark specks in all cells	Gravity

Activity

- Get the students to draw lines linking each early idea or discovery to the modern explanation or use.

Answers

Beads formed by fire in desert areas	Glass
A falling apple	Gravity
The Sun rising and setting	The Earth spinning on its axis
Bird wing shape	The aeroplane
Pollen grains vibrating on water	Brownian motion and diffusion
Fossils	Evolution
Dark specks in all cells	The nucleus

Differentiation

- ⇑ More able students could recall the simple observations from investigations that they have carried out which have then led to more complicated scientific explanations.

Dividing duckweed

Objective covered/aim

Framework SE2.

Resources

Poster/sugar paper. Chunky pens.

Activity

- Read this problem to the class:

  Ben has a problem. His pond has a tiny plant called duckweed growing on its surface. He keeps removing it, but it keeps growing and covering the surface of the water. Can you help him find out how to stop it growing?

- The students should work in pairs. Ask them to write their answers to these questions on poster paper:

  1 What factors may be causing the plant to grow?
  2 Ben thinks as the pond is in a sunny position the plant may like this. How could you investigate this to see if he is right?
  3 What factors should be kept the same to make it a fair test?
  4 Do you think the plant will grow faster in the light or dark?

Answers

1 Water temperature, amount of light, amount of minerals in the water, amount of fertiliser in the water, lack of any organism which eats the duckweed.
2 Put a single plant in a beaker of water. Put in a sunny position. Repeat with a plant in a shady position, and one in the dark.
3 Same volume of water, same temperature of water, same species of plant, same room temperature, same water.
4 The plant will grow faster in a sunny position, as plants need light to make food. More food leads to more energy for growth.

Ben's beans

YEAR 7

Objective covered

Framework SE3.

Aim

To practise writing methods in a point-by-point style.

Activity

- Ask the students to write a method for making baked beans on toast. Encourage them to write the method in a point-by-point style. For example, they might start:

  1 Carefully open the tin using a tin opener.

- Ask what could go wrong when making baked beans on toast. 'How could you control this?'

Differentiation

- ⇑ Encourage the students not to use the word 'I'.
- ⇑ Encourage the students to consider alternative methods, such as cooking in a microwave.
- ⇓ Encourage the students to show their method using pictures.

Know your tools

YEAR 7

Objective covered

Framework SE4.

Aim

To recall the equipment and units used in science.

Activity

- Ask the students what equipment they would use to measure the following:
  1. The length of the lab.
  2. The temperature of tea.
  3. The mass of an object.
  4. The volume of water.
  5. The weight of an object.
- Go on to ask what units they would use for these measurements.

Answers

1 Metre rule, metres (m)
2 Thermometer, degrees Celsius (°C)
3 Top pan balance, grams/kilograms (g, kg)
4 Measuring cylinder, cubic centimetres (cm^3)
5 Newton meter, newtons (N)

Differentiation

- ⇑ The students could consider the use of ICT, for example data-loggers.
- ⇓ Give less able students pictures of scientific equipment to label.

Millie's measurements

Objective covered/aim

Framework SE5.

Resources

OHP and OHT, board or worksheet prepared with the following:

Millie bought four new tyres for her little red car. After three days the car felt odd. She took it back to the garage where the mechanic tested all the tyres five times. These are the results

	Front right	Front left	Rear right	Rear left
Pressure (in pounds)	28	28	26	27
	28	28	28	28
	28	27	27	26
	28	28	26	27
	28	28	26	27

Activity

- Ask the students to use their knowledge to answer these questions:

  1 Which reading is the most frequent for each tyre?
  2 What is the average reading for each tyre? Round up to nearest whole number.
  3 Is it the tyres that are making the car feel 'odd'? Justify your answer.
  4 Can you imagine anything else that may make the car feel 'odd'?

Answers

1 28, 28, 26, 27
2 28, 28, 27, 27
3 By outcome.
4 Shock absorbers/wheels need balancing/slow puncture.

Differentiation

- ⇓ Get the students to draw Millie and her car, labelling the tyres with the different pressures.

What grows in Emily's lawn?

YEAR 7

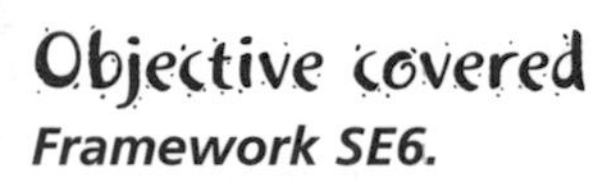

Objective covered

Framework SE6.

Aim

To be able to draw graphs of experimental results and interpret them.

Resources

Graph paper, rulers and colouring pencils.

Activity

- Tell the class that Emily wants to know what plants grow in her lawn and which are the most common.
- Display the following table:

Plant	Number found
Clover	12
Buttercup	4
Daisy	6
Dandelion	7
Plantain	8

- Tell the students that Emily used a quadrat to get the results in the table. (A quadrat is a small area of ground marked off for the detailed investigation of animal or plant life.)
- Ask them to help her see what's going on by drawing a bar graph. They should colour the most common plant in red and the least common in blue.

Differentiation

- ⇑ Ask the students why a line graph is not appropriate.
- ⇑ They could present the results as a pie chart.
- ⇓ Give students graph paper with axes already drawn for them.

Graphic graphs

Objective covered

Framework SE7.

Aim

To recognise that data may produce a variety of shapes in a graph.

Resources

Prepare an OHT, board or worksheet with the following statements and graphs:

1 Taller trees have thicker stems.
2 Older cars are less efficient.
3 Gravity does not change with height.
4 Reaction rates increase with temperature.
5 Older teachers have slower reaction times.

a

b

c

Activity

- Ask the students to link each statement to the appropriate graph.

Answers

1 c 2 a 3 b 4 c 5 a

Differentiation

- ⇑ Do not provide the graphs – only the statements. Ask the students to draw a sketch graph for the statement.
- ⇓ Less able students could be given the questions and graphs on a worksheet.

What does it all mean?

Objective covered

Framework SE7.

Aim

To practise writing simple conclusions to graphs.

Resources

OHP and OHT, board or worksheet prepared with simple graphs. Here are two ideas:

A graph to show how 100 cm^3 of boiled water cools over time

Temp (°C)

Time (min)

A graph to show how quickly a crystal grows in a saturated solution

Size of crystal (mm)

Time (days)

Activity

- Ask the students to write simple conclusions to the graphs.

Differentiation

- ⇑ The students could write a conclusion to a more complicated graph, for example:

Bubbles per minute

Light intensity

The number of bubbles released per minute from pond weed as the intensity of light is increased

Solid as a rock

YEAR 7

Objective covered/aim

Framework SE8.

Resources

OHP and OHT, board or worksheet prepared with the following:

PC Kendall reported to the sergeant on duty that four witnesses had given him this evidence about the burglar who broke into the Science Emporium the evening before.

	Colour of hat	Colour of hair	Height	Body type	Colour of jacket
Witness 1	Black	Dark	5'10"	Large	White
Witness 2	Dark	Fair	6'0"	Slim	Dark
Witness 3	Navy	?	Short	Medium	Dark
Witness 4	Red	Dark	Tall	Big	Black

Activity

- Ask the students the following questions:

  1 What can you conclude about the burglar?
  2 Is this evidence strong enough? Justify your answer.
  3 What further information might you need?
  4 Where could you get this information?

Answers

1 Probably had a darkish hat, dark hair, is tallish, biggish build, had darkish jacket.
2 No – too many variables.
3 By outcome.
4 Video/CCTV/reconstruction of the crime/more witness statements.

Differentiation

- ⇑ Ask these additional questions:

  5 What do we call the science of investigating crimes? (Forensics)
  6 Imagine you are an investigator. What would you look for at this crime scene?

- ⇓ Create a 'Wanted' poster of this villain.

New ideas

YEAR 8

Objective covered

Framework SE1.

Aim

To recall that accepted scientific ideas and concepts can evolve over time.

Resources

OHP and OHT, board or worksheet prepared with the following:

Over the years scientific concepts and ideas have changed. From where did these scientific concepts come, or into what did they evolve?

1 The world is flat → ?
2 The plague was carried in the smell in the air → ?
3 ? → Darwin's theory of evolution
4 The Earth is in the centre of the Universe → ?
5 Mother nature would never allow a vacuum to be created → ?
6 You can make gold from other metals → ?
7 The four elements are fire, earth, wind and water → ?
8 The Moon is made of cheese → ?

Activity

- Get the students to fill in the question marks.

Answers

1 It is roundish (geoid)
2 Fleas on rats carried it
3 God created all living things as we see them
4 The Sun is in the centre of our solar system
5 We can create a vacuum
6 Alchemy has never succeeded
7 There are many chemical elements
8 It is made from rock

Differentiation

- ⇑ More able students could consider whether any current scientific ideas could change or be completely disproved.

In hot water

Objective covered

Framework SE2.

Aim

To recall that methods should be designed to give the most accurate results possible.

Resources

OHP and OHT, board or worksheet prepared with the following:

A class is asked to investigate the best way of insulating a cup. Veronica writes the following method:

1. Fill a cup with hot water.
2. Insulate it.
3. Put in a thermometer.
4. Measure the temperature 5 minutes later.

Activity

- Ask the students to explain why this is not a good method.
- Ask them to write an improved method, which will yield more accurate results.

Answers

The students could improve the method by:

- stating what type of cup they will use
- stating how they will insulate the cup, perhaps by using a lid or types of insulation
- having a uniform start temperature
- measuring the temperature with a data-logger to improve accuracy
- doing more repetitions
- running the experiment for a longer period of time.

Differentiation

- ⇑ The students could explain why their investigation would be a fair test.
- ⇓ The students could write or draw their list of apparatus.

Information please

Objective covered

Framework SE3.

Aim

To improve awareness of secondary sources of information.

Activity

- Tell the students to use their knowledge and experience to answer these questions:
  1 Where can you get scientific information?
  2 a Which secondary source of information would be best when planning an investigation?
    b Which secondary source of information would be best when writing a conclusion?
    c Which secondary source of information would be best when looking for the most up to date information?
    d Which source of secondary information would be best when looking up more general information?

Answers

1 Internet, encyclopaedia (books or CD-ROM), textbooks, newspapers and magazines, library and so on.
2 By outcome. Suitable answers include:
  a Textbook
  b More advanced textbook
  c Internet, newspapers and magazines
  d Encyclopaedia

Differentiation

- ⇑ Ask the students under which Dewey numbers in the library scientific textbooks are found. (500–599)

Yes Sah!

Objective covered

Framework SE4.

Aim

To collect simple data and process them quickly in a variety of ways.

Resources

OHP and OHT, board or worksheet prepared with the following:

Name	**Shpetim**	**Elijah**	**Abida**	**Joseph**	**Ben**
Height (cm)	167	154	162	172	160
Weight (kg)	64	48	51	70	61

Activity

- Ask the students to arrange the data in order, from tallest to shortest.
- Repeat, this time asking the students to order the data from heaviest to lightest.
- Tell the students to present the weight data and height data as two separate bar charts.

Answers

- Tallest to shortest: Joseph, Shpetim, Abida, Ben, Elijah
- Heaviest to lightest: Joseph, Shpetim, Ben, Abida, Elijah

Differentiation

- ⇑ Ask the students to produce line graphs and identify the differences between them.
- ⇓ Just ask the students to order the data or draw the bar charts.

Have a swinging time!

Objective covered

Framework SE5.

Aim

To plan an enquiry to find out what affects the swing of a pendulum.

Resources

OHP and OHT, board or worksheet prepared with the following:

> The seventeenth-century scientist and philosopher Galileo attended church and watched the huge light fitting gently swing. He timed how long it took to swing left to right and back again. He thought, 'Does the swing time change if the chain length changes, or if the mass of the light changes?'

Activity

- Get the students to plan an experiment to provide data that are accurate and would impress Galileo.

Differentiation

- ⇑ The students could use their ideas to write a prediction and then research the historical outcome of the experiment.
- ⇓ Get the students to produce a simple picture showing why the experiment was thought of in the first place. Use this to emphasise that observations can lead to great scientific discoveries.

Sort yourself out

Objective covered

Framework SE6.

Aim

To practise gathering and analysing data.

Activity

- Ask the students to write down their date of birth, hair colour and height.
- Each student then shares their data with four others so that each student has five dates of birth, five colours of hair and five heights.
- Tell the students to order their results from oldest to youngest, tallest to shortest and darkest to fairest.
- See if the students can spot a pattern or relationship between age and height.
- Ask the students what conclusions they can draw about the five students whose data they have looked at.

Differentiation

- ⇑ More able students can discuss whether this is a fair test. They should explain and justify their answer.
- ⇓ Give the students these sentences to fill in:

  The oldest is __________ tall and has _______________ hair.

  The youngest is __________ tall and has ______________ hair.

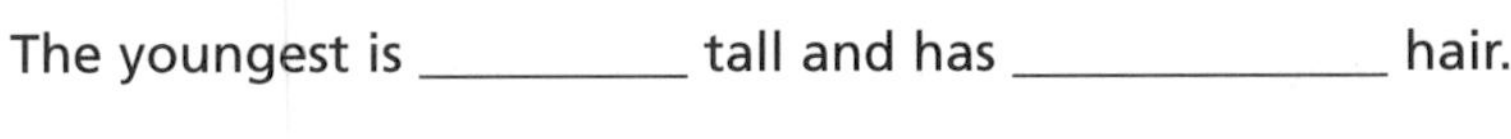

Funny results

Objective covered

QCA 8I. ***Framework SE7.***

Aim

To be able to recognise anomalous results and recall how to reduce the likelihood of their occurrence.

Resources

OHP and OHT, board or worksheet prepared with the following:

A group of students did a single experiment to see how 100 cm^3 of hot water would cool over a 5-minute period of time.

Time (min)	Temperature (°C)
0	100
1	92
2	84
3	76
4	79
5	60
6	52

1 What is meant by the term an 'anomalous result'?
2 Is there an anomalous result in this set of data?
3 How could you improve the experiment to reduce the effect of anomalous results?

Activity

- Tell the students to answer the questions in their books.

Answers

1 A result that does not fit the pattern of the rest of the results.
2 Yes – at 4 minutes.
3 Do more repetitions.

Differentiation

- ⇑ Ask the students to draw a graph of the data.
- ⇓ The definition of 'anomalous' may need to be given.
- ⇓ The students may need to be given a graph of the data to help them answer question 2.

Stem cells

Objective covered/aim

QCA 9A. ***Framework SE1.***

Resources

OHP and OHT, board or worksheet prepared with the following statements:

1 In the future, new nerves for spinal injuries may be grown from stem cells.
2 People with damaged spines could walk again.
3 Stem cells have the ability to become any tissue or organ.
4 If one of your organs failed you could have a new one grown for you.
5 Organs grown would not be rejected.
6 It involves cloning you.
7 The embryo grown to supply the stem cells is later destroyed.
8 The clone of you will be born and you will have to look after it.
9 The cloned embryo that supplies the stem cells is identical to you.
10 The cloned baby is never born.
11 Many people are against the disposal of human embryos.
12 There would be no need for organ donors.

Activity

- Remind the students that stem cells are special cells, which can grow into any tissue or organ. A growing foetus has lots of stem cells. Stem cell technology has some very exciting possibilities for the future and some ethical problems too.
- Ask the students to discuss whether the statements about stem cell technology are true or false.

Answers

1 True	2 True	3 True	4 True	5 True	6 True
7 True	8 False	9 True	10 True	11 True	12 True

Differentiation

- ⇑ 'Who owns your clone? Is it you as it's identical to you, or your parents whose genes it has?'

New technology

Objective covered

Framework SE1.

Aim

To recall that scientific technology is a rapidly evolving area.

Resources

OHP and OHT, board or worksheet prepared with the following:

Over the years, scientific technology has changed. Where did these modern technologies come from or what have these old pieces of equipment become?

1 Abacus → ?
2 ? → Electron microscope
3 Paraffin burners → ?
4 Mercury thermometers → ?
5 Pen and paper → ?
6 ? → CD-ROM
7 Letters and the postal service → ?
8 Inch, mile, pounds and ounces → ?

Activity

- Get the students to fill in the gaps.

Answers

1 Calculator
2 Light microscopes
3 Bunsen burners
4 Data-loggers or electronic thermometers
5 Computer
6 Reference books
7 e-mail
8 cm, km, kg and g (the SI units)

Differentiation

- ⇑ Ask the students to consider the pros and cons of these new technologies. Examples: Electronic thermometers can be more accurate but are much more expensive. The use of SI units makes the transfer of ideas easier – until they have to be converted. This has led to mistakes being made when Europe and America have worked together on space projects.

Plan it out yourself

Objective covered

Framework SE2.

Aim

To recall that different problems require different strategies to solve them.

Resources

OHP and OHT, board or worksheet prepared with the following questions:

- Does tea cool down faster if the milk is added immediately after brewing or if it is added 10 minutes later?
- Will more salt or more sugar dissolve into 100 cm^3 water?
- Does the pH of rainwater change with the seasons?
- Do low energy light bulbs really save money?
- Is there more bacteria on your hands before or after you wash your hands?
- Does maximum heart rate increase or decrease with age?

Activity

- Ask the students to choose two questions to investigate and to create methods by which to investigate them.

Differentiation

- ⇑ The students should write full point-by-point methods for their chosen investigations.
- ⇓ The students may require a writing frame.

Random results

YEAR 9

Objective covered

Framework SE3.

Aim

To recall that trial runs should be made in order to make refinements and adjustments.

Resources

OHP and OHT, board or worksheet prepared with the following:

The following method was carried out to see how temperature affects the rate of dissolving.

1. Heat water to desired temperature.
2. Add sugar.
3. Time how long it takes to dissolve.

The following results were obtained:

Time to dissolve (s)

Temp (°C)

Activity

1. How do you think temperature affects the rate of dissolving?
2. Do you agree with the graph of results obtained?
3. Create a new method to obtain more accurate results.

Answers

1. The higher the temperature the faster the sugar dissolves.
2. No
3. Dissolve the same amount of sugar in the same volume of water, always stirring or not stirring, using the same equipment, at specified temperatures. Do the same number of repetitions at each temperature.

Differentiation

⇑ Tell the students to justify their new methods.

⇓ Ask the students to draw the equipment they would use in this experiment.

Get the right graph

Objective covered

Framework SE5.

Aim

To practise using the correct type of graph for a given set of data.

Resources

OHP and OHT, board or worksheet prepared with the following questions:

1 List as many different types of graph as you can.

2 State which type of graph you would draw for each of these graph titles:

- a A graph to show the frequency of different colours of cars in a car park
- b A graph to show how the height of cress changes as it grows from seed, over a 10-day period
- c A graph to show the number of boys and girls in a class
- d A graph to show the cooling of 100 cm^3 of hot water over a 10-minute period

Activity

- Get the students to use their knowledge to answer the questions.

Answers

2 a Bar chart or pie chart
- b Line graph
- c Bar chart
- d Line graph

Differentiation

- ⇑ Get the students to draw a quick sketch graph to show how they think the axes of the graphs should be labelled.
- ⇓ Less able students could draw examples of each type of graph.

Explain the changes

YEAR 9

Objective covered

Framework SE6.

Aim

To recall that scientific knowledge can be used to explain and interpret patterns.

Resources

OHP and OHT, board or worksheet prepared with the following:

These average temperatures were recorded over a year at Cross Lanes Farm:

Month	Temperature (°C)
Jan	5
Feb	4
Mar	6
Apr	8
May	13
Jun	20
Jul	24
Aug	26
Sep	19
Oct	14
Nov	9
Dec	7

Key words: orbit, tilt, axis, towards, seasons, Sun, energy, away, rotation

Activity

- Ask the students to use their scientific knowledge to explain the pattern in temperature variations.
- Encourage them to use the key words.

Differentiation

- ⇓ Less able students will need a writing frame.

Firm conclusions

YEAR 9

Objective covered

Framework SE7.

Aim

To recall ways of making investigations fairer and more accurate.

Resources

OHP and OHT, board or worksheet prepared with the following:

Victor carried out an experiment to see how the size of a beaker affects the rate of boiling of water. Here are his results:

Volume of water	Time to boil
50 cm^3	5 min
100 cm^3	6 min
150 cm^3	5 min

Activity

- Tell the students to look at the results then ask:
  1. Can you make a firm conclusion from these results?
  2. How could you improve the method to make a firm conclusion?

Answers

1. Firm conclusions cannot be drawn from these results.
2. For example: repetitions, timing in seconds, use of electronic thermometers or data-loggers, five different volumes of water

Differentiation

- ⇑ More able students could redesign the experiment to make it a fair test.
- ⇓ Less able students could try to predict the outcome of the experiment and justify their answers.

Published by Letts Educational
The Chiswick Centre
414 Chiswick High Road
London W4 5TF
☎ 020 89963333
020 87428390
✉ mail@lettsed.co.uk
www.letts-education.com

Letts Educational is part of the Granada Learning Group. Granada Learning is a division of Granada plc.

First published 2002

ISBN 1 84085 4618

British Library Cataloguing in Publication Data
A catalogue record for this book is available from the British Library.

Commissioned by Helen Clark
Project management by Vicky Butt
Editing by June Hall and Mark Haslam
Cover design by Ken Vail Graphic Design
Internal design by IFA Design Ltd
Illustrations by IFA Design Ltd
Production by PDQ
Printed and bound by Ashford Colour Press